Miracle on Fourth Street

Miracle on Fourth Street

Saving an Old Merchant's House

Mary L. Knapp

Girandole Books, New York
The Merchant's House Museum, New York

Published by Girandole Books and the Merchant's House Museum
590 West End Avenue, New York, New York 10024

Designed by Steven M. Alper

Printed in the United States of America

Publisher's Cataloging-in-Publication data
Names: Knapp, Mary, 1931-
Title: Miracle on Fourth Street : saving an old merchant's house / Mary L. Knapp.
Identifiers: ISBN 978 0997164626 | LCCN 2015960135
Description: First edition | New York [New York]: Girandole Books and Merchant's House Museum, 2016. | Includes bibliographical references and index.
Subjects: LCSH Merchant's House Museum (New York, N. Y.)--History. | Historic buildings--New York (N.Y.) | Historic buildings--Conservation and restoration--New York (State)--New York. | Buildings--Repair and reconstruction--New York (State)--New York. | Architecture--Conservation and restoration--United States--History--19th century. | Historic preservation--New York (State)--New York. | BISAC ARCHITECTURE/Buildings, Landmarks and Monuments. | ARCHITECTURE / Historic Preservation/Restoration Techniques. | ARCHITECTURE /Historic Preservation/General.

Classification: LCC F128.37 .K54 2016 | DCC 069/.53—dc23

For all of the volunteers who through the years have helped tell the story of the Tredwell family and their house.

A building represents the height of achievement of the people of an age. Pictures can never preserve the spirit of the building; you have to be able to feel the space around you.

JOSEPH ROBERTO

You should respect the treasure.

SARAH TOMERLIN LEE

Table of Contents

Introduction

Seabury Tredwell came of age along with his country. He was born in 1780, one year before the British surrendered at Yorktown, and he died in 1865, just one month before the end of the Civil War.

In 1798, at the tender age of 18, he left his rural Long Island home for New York City where, like so many other young men, he intended to make his fortune in the maritime trade. After five years as a clerk, he established his own hardware importing business on Pearl Street in 1803. Seabury remained in business for 32 years, 20 of them in partnership with two of his nephews. It was an extraordinary era when the "Old Merchants" laid the commercial foundations of New York City.

During those years, sailing ships brought thousands of barrels and kegs filled with screws and nails, spittoons and buckets, hammers and rakes, knives and forks, buttons and thimbles, and even pistols to the hardware business of Tredwell and Kissam.

Like the other Old Merchants during the early years of the nineteenth century, Seabury lived at the tip of the island of Manhattan in close proximity to the harbor. But as the volume of commerce exploded during the years following the War of 1812, the area became crowded with lofts and warehouses. Bales and barrels blocked the sidewalks; laborers shouted and cursed as they loaded and unloaded cargo. Delivery wagons clattered along the cobblestones.

And so in 1820, a suburb began to materialize as the Old Merchants, seeking a more hospitable environment for their families, began the exodus from the waterfront, building new homes in the Bond Street Area, just east of Washington Square. In 1835, the Tredwell family finally joined their neighbors, making the move uptown to Fourth Street—and there they stayed.

When in the 1840s the neighbors began moving further uptown, building even more elegant houses on Fifth Avenue and around Gramercy Park, the Tredwells continued to live in the gradually declining Fourth Street neighborhood. After Seabury died in 1865, Eliza Tredwell and the couple's four unmarried daughters lived out their lives in the Fourth Street house.

Certainly the Tredwells did not intend to preserve their house for posterity. They were simply natural conservationists. In the nearly 100 years of their residency, they never remodeled their home, never installed a central furnace, never embraced the nineteenth-century craze for wallpaper, never fully wired the house for electricity. They stopped buying new furniture sometime in the 1850s and before that, when they did, they saved the old, simply moving it to another part of the house or carting it off to their farm in New Jersey. The women of the family gently folded their outmoded dresses and carefully packed them away in trunks.

It was not only the family's innate conservatism that served to preserve the house but the fortuitous appearance of four dedicated individuals at crucial periods in the house's history and the passage of the New York City landmarks legislation in 1965, which provided legal protection against demolition of the property.

After the death of the last Tredwell sister, Gertrude, George Chapman, a grandnephew of Mrs. Tredwell, saved the house and its contents from the auction block and established a nonprofit organization to run the house as a museum. In 1962, Cornelia Van Siclen, a member of The Decorators Club, persuaded her organization to take the house on as a project. The club kept it going during some of the darkest days of its history, and continued to support it for over 25 years. When the house was on the verge of collapse in the 1970s, Joseph Roberto, New York University staff architect, undertook a total structural restoration, giving the house a new lease on life. After Roberto's unexpected death, Margaret Halsey Gardiner became the director in 1990 and immediately confronted a crisis caused by the reckless demolition of the adjacent building that threatened the very existence of the house. She enlisted the support of preservationists, private foundations, government agencies, and volunteers, placing the museum on a professional footing, and insuring its financial stability.

The Tredwells did not anticipate the posthumous scrutiny they and their possessions have received since 1936 when their house was opened to the public. One suspects that the entire family—from Seabury on down— would be horrified at strangers walking through their home, studying their furniture, their clothing, and their candlesticks, commenting on their taste. But perhaps if they realized what a unique opportunity they left us for understanding the past, they would forgive us for being so nosy.

To feel the world of the nineteenth century in our bones, it's necessary to find a *place* that can take us there. The Merchant's House is just such a place. It tells the story of just one family, a family who left us enough of their possessions, so that when we move through the house, we not only retrace their steps but we see what they saw. The objects around us felt the touch of the Tredwells, and though we are not allowed to touch these things today, we can easily imagine what it was like to play their piano, chop vegetables on the scarred kitchen worktable, or sit down to a formal dinner on one of their Duncan Phyfe chairs. We look into the mirrors that once reflected their faces, and when we're tired, we long to lie down in their beds.

Once I was giving a tour of the house to second graders. The children were seated on the parlor carpet in front of me. I explained that a family with eight children lived in the house over 100 years ago and today the house was still here just as it had been then. The furniture was theirs; the big sister played the piano; the family sat on the chairs.

A hand shot into the air. The seven-year-old's eyes were wide. He pointed to the center table.

"You mean ... you mean ... that's the *really real* table?"

Yes, that's it. The Merchant's House is real. That's why so many love it.

Mary Knapp
New York City, 2016

1. The Old Merchants House and Its Nineteenth-Century Neighbor, 1937

New York City Municipal Archives

By 1937, the elegant row of nineteenth-century homes on East Fourth Street had all but disappeared, and the neighborhood consisted largely of garages, warehouses, and manufacturing lofts. The house shown on the west (left) of the Old Merchants House was torn down in 1947 and replaced with a one-story building.

1

1933–35
A Bold Idea

In 1840, when Gertrude Tredwell was born in the red brick and white marble rowhouse on Fourth Street, the neighborhood was a quiet leafy suburb where the Old Merchants of New York City enjoyed the good life. At the time of her death in 1933, almost a century later, the country was in the grip of the Great Depression. The trees were gone; factories and warehouses had replaced most of the three- and four-story brick rowhouses, and the tracks of the Third Avenue Elevated Railway cast a gloomy shadow over the nearby once-fashionable Bowery.

For several years before she died, Gertrude was confined to her second-floor bedroom. As a girl, she danced the quadrille in the parlors that were now shuttered, dusty, and dark. The only heat in her room came from the fireplace; newspapers stuffed behind the shutters helped to keep out the cold. A paid companion looked after her personal needs, and a resident caretaker couple took care of the house. She had few visitors, for she had outlived most of her family. Only her nieces, Lillie Nichols and Betty Stebbins, and her sister Mary Adelaide's grandson, Charles Van Nostrand, came to see her. She had also outlived her father's substantial fortune. As the money ran out, she survived over the years by selling the 850-acre family estate in Rumson, New Jersey, piece by piece, and by borrowing from the Bank for Savings and from Lillie Nichols. The bank held a first mortgage of $14,000; Lillie held a second mortgage of $7,000 and a third mortgage of $4,000. Lillie Nichols was Gertrude's heir; the heavily mortgaged house, her legacy.

As executor of Gertrude's estate, Lillie prepared to sell the house and its contents at auction. Finding a buyer for a 100-year-old house in a derelict neighborhood was not going to be easy. And yet, Charles Van Nostrand, who had an emotional attachment to his Great-Aunt Gertrude and to the Tredwell home, knew someone who just might be interested.

George Chapman was a 63-year-old attorney with an apartment on Park Avenue, a country home called “Airlie” in Mt. Kisco, New York, and a patrician attitude. His parents were cultured American expatriates, members of the American community in Florence, Italy, where he was born in 1870. He attended St. Paul’s School, in Concord, New Hampshire; graduated from Harvard University, class of 1892; and attended New York Law School. Politically conservative, Episcopalian by faith, he was married to Beatrice Wright, who was distantly related to Alexander Hamilton. The couple was listed in the social register. They had no children.

Chapman’s grandmother and Gertrude Tredwell’s mother were sisters, which made Gertrude his first cousin once removed. Although he was not particularly close to the Tredwells, Chapman had visited their home on occasion, and he kept in touch with his young Tredwell relative, Charles Van Nostrand.

2. Geoge Chapman, 1892
HUD 292.50 Harvard University Archives
This photograph of George Chapman, the founder of the Old Merchants House, was taken in 1892 when he graduated from Harvard University. Forty-one years later, he would acquire the Tredwell home and begin the process of establishing a historic house museum.

Charles believed saving the house from the auction block was something that just might appeal to Chapman, knowing that he had an extreme fondness for things antiquarian. As president of the Fifth Avenue Building Corporation, Chapman had been on hand in 1908 for the demolition of the celebrated 1859 Fifth Avenue Hotel on Twenty-Third Street where the Fifth Avenue Building would be built. He had rescued the white marble mantel from the drawing room of the hotel himself and had it installed in his office, which was hung with photographs, prints, and paintings of the city when it was still a colonial seaport.

On the day before the sale, the auctioneers were busy getting ready. It was a dismal prospect, as estate sales tend to be. For the first time in years, the pocket doors of the double parlor were thrown open, and the accumulation of decades lay on the auctioneer's tables awaiting the cool inspection of strangers. The Tredwell women's dresses, once so fondly fussed with, were bundled like rags. The shabby and worn furniture was in the process of being tagged. When Charles learned the sale was imminent, he wasted no time in contacting Chapman.

Years later, in an interview with Helen Erskine, who was preparing to write a book on New York City recluses,[1] including Gertrude Tredwell, Chapman himself gave her this account of how he came to buy Gertrude's house:

> "At the eleventh hour, Charley Van Nostrand appealed to me to rescue the place," said Mr. Chapman. "The sale was scheduled for the very next day, when he dashed into my office. 'George,' he said, 'you've got to save the East Fourth Street house. Cousin Lillie's auctioning off everything tomorrow.'" Mr. Chapman smiled. "I'm a sucker for old houses. Being foolish, I said, 'Okay Charley, we'll save it.'"

The two men grabbed a cab and raced down to East Fourth Street.

> The auctioneer's men were at work, tying up rare old cashmere shawls in bundles, ticketing the Duncan Phyfe parlor furniture. "It was quite dramatic, as I recall it," remarked Mr. Chapman. "I ordered the men to stop work. They replied that they could not without instructions from their boss. I long-distanced Lillie at her summer home in Connecticut, made her an offer for the house and contents. She accepted it and promised to telephone the auctioneer."

---

1 Erskine, Helen Worden, *Out of This World* (New York: Putnam, 1953), 170–71.

We don't know exactly what Chapman said to Lillie that day to persuade her to postpone the sale, but it would be a little over a year before he came up with a detailed proposal. The first thing he did was to call on two of his personal friends who could offer an expert opinion on the historic worth of the property: Benjamin Wistar Morris, a noted architect and a member of the board of directors of the Metropolitan Museum of Art, and Hardinge Scholle, director of the Museum of the City of New York. Chapman told Helen Erksine that his friends were impressed: "After exploring all the rooms, Mr. Morris sat down on one of the chairs in the hall. 'George,' he said, 'this house is 100 percent authentic. You've got a gem.'" Scholle agreed that in spite of its deteriorated condition, the house was worthy of preservation, particularly since it was still furnished with the original owner's period furniture. It was then that Chapman made his final decision to buy the house and to convert it to a historic house museum. Hardinge Scholle helped Chapman devise a plan that would realize those aims, and on November 2, 1934, they presented a proposal to Lillie's attorney.

Chapman planned to set up a nonprofit corporation to be called the Historic Landmark Society. This newly formed corporation would assume the indebtedness, take title to the house, and open it to the public as a museum. Chapman convinced Lillie Nichols to cancel her $4,000 third mortgage altogether and to reduce the $7,000 second mortgage to $4,000, which he offered to repay at five percent annually for five years. He also personally offered to pay her $1,000 for the contents of the house, which he would then lend to the society. If for some reason things did not work out and the bank was forced to foreclose, Chapman would still own the chattels.

It was a proposal Lillie could hardly refuse under the circumstances. She was then 79 years of age, in frail health, and likely eager to have the matter of Gertrude's estate off her hands. And, as holder of the second and third mortgages, she probably would not have realized a return from the auction of the house if she had gone ahead with it. So on December 7, 1934, Lillie agreed to these terms. Thus she would receive a tidy sum for the contents of the house and an annual stipend for five years to come. And Chapman would be able to acquire the property quietly without any immediate expenditure of cash nor the involvement of others. That is the way he wanted it. There would be no committees to deal with, no drawn-out efforts to raise money to purchase the house, and no one to second-guess his decisions.

Chapman did not have a structural appraisal done to determine just how much it might cost him to put the house in shape. Nor did he have a very clear idea of how much it would cost to maintain a house of this vintage. If ever the aphorism, "Fools rush in where angels fear to tread," was apt, it was here. But he had made up his mind to preserve the house as a museum, and he would stick by his decision until his death in 1959. There was never a question of where the corporation would get the money to make the mortgage payments or where the inevitable shortfall between the small admission fee he intended to charge and the maintenance costs would come from. Chapman was prepared to take up the slack himself.

3. **Helen Erskine and Dame Judith Anderson**
Helen Erskine (right), chats with the actress, Judith Anderson, at a reception held at the Old Merchants House on September 24, 1953, to celebrate the publication of the book, *Out of This World*.

George Chapman's desire to save the Tredwell house was more than his simply being "a sucker for old houses," as he had told Helen Erskine. On May 12, 1937, he wrote to John Vedder, vice president of The Bank of New York, asking for his help in researching Seabury Tredwell's hardware firm. He explained his motive for restoring the Tredwell home:

> One of the ideas behind the restoration of the Tredwell house ... was that New York after all is preeminently a commercial city, and that whatever of greatness it possesses aside from its natural advantages has been created by its merchants and businessmen, many of whom came to the city from outside ... and made their way to success and contributed to the growth of New York. Seabury Tredwell was to some extent typical of such men ... and it seemed fitting that his home should be preserved not only as a fine example of a dwelling of the early days, but as a memorial to a merchant and one of the class and type through whose constructive efforts the commercial supremacy of New York has been brought about.

In saving a house that had belonged to an early New York City merchant, Chapman intended to memorialize a social class he believed embodied the fundamental virtues of American life, in his view a class vastly superior to the vulgar opportunists who were attracted to the city in the pursuit of wealth and power. It was this desire to memorialize the early merchants that led him to name the museum the Old Merchants House. It also sometimes led him to misrepresent the house as being of an earlier vintage than it actually was.

It was not until 1936, when the Department of the Interior sent a representative to research the history of the house, that the exact date of its building, 1831–32, was established.[2] Chapman had assumed it had been built earlier, and even after he knew better, he would often mistakenly claim it had been built "around 1826" and that the probable architect was John McComb, who along with Joseph Mangin designed the New York City Hall in 1811. For years he tried unsuccessfully to discover evidence McComb designed the house.

Chapman always referred to the architectural style of the house as "purely

2 The historic report was written as part of a government program designed to give out-of-work artists and writers employment during the Great Depression.

Federal," the building style popular in New York City from the Revolutionary War to the 1830s. The fact was that by 1832 when the Tredwell home was built, many of the Old Merchants had abandoned their Federal-style homes in lower Manhattan, where they lived cheek by jowl with their businesses, and had moved north to the residential suburbs of Greenwich Village and the Bond Street Area. There some of them built houses in the newly fashionable Greek Revival style, like the buildings that can still be seen on the north side of Washington Square. Other houses in the neighborhood were built in a transitional style, a hybrid of the Late Federal and Greek Revival. The Tredwell house was one of these houses. Built of red brick, with a steeply pitched roof, dormer windows, and a fan light over the door, the house had a primarily Late Federal façade, but inside it was pure Greek Revival.

However, since Chapman saw his restoration effort as a way of preserving a house that had belonged to one of the Protestant merchant elite of the early Republic, that is to say a "purely Federal" house, he simply ignored evidence of the later Greek Revival style. In October 1943, when the Metropolitan Museum of Art included pictures of the double parlor in its exhibition on Greek Revival architecture, he complained that "they suggested a cat in a strange garret." He was not happy to have Talbot Hamlin include the Old Merchants House in his definitive work, *Greek Revival Architecture in America* in 1944. And when *Town and Country* magazine wanted to include the house in a series on Greek Revival architecture in 1948, he wrote a memo directing Beatrice Clark, his personal secretary, to turn them down, reminding her, "We are Federal, period."

Had the Tredwell house truly been a "purely Federal" building, it would have fit Chapman's plans perfectly. Restoring such a house would have been more in keeping with the work of William Sumner Appleton, and Appleton was someone whom Chapman deeply admired and wanted to emulate. A few years older than Chapman, he was also educated at St. Paul's and Harvard, and in 1905 at the age of 31 he was one of three persons who led the effort to save and then to completely restore the house of Paul Revere in Boston. Sumner Appleton went on to found the Society for the Preservation of New England Antiquities (SPNEA)[3] in 1910 and to become a towering figure in the field of historic preservation. The mission of SPNEA was, in Appleton's words, "to save for future generations structures of the seventeenth, eighteenth, and the early years of the nineteenth century, which are architecturally beautiful or unique or have special history."

---

3 The name of the society was changed to Historic New England in June 2004.

At first, Chapman speculated that someday his nonprofit society might acquire other properties and do for New York what Appleton was doing for New England, but he soon abandoned this idea.

Once Chapman had settled the terms of the transfer of the property with Lillie Nichols, he set about obtaining a charter for the Historic Landmark Society, Inc., to be organized under the membership corporation law of the State of New York. The certificate of incorporation was finally filed on January 28, 1935. The society was granted the right "to acquire and maintain as a museum or museums, one or more parcels of real estate with the buildings thereon, having historic, educational and/or literary interest," with the purpose of promoting interest in the "study of landmarks, buildings associated with historic or literary events, characters, and/or the manner of living of earlier generations."

The bylaws of the new society provided for a three-member board of directors. Chapman, of course, would serve as president, and Hardinge Scholle agreed to serve as vice president. The third member was Nat Tyler, a business colleague of Chapman. Tyler resigned the following year and was replaced by John Dix, a friend of Chapman who lived in Mt. Kisco and was a vestryman of Trinity Church. In 1937, a provision to increase the number of trustees to as many as five was put into effect, and Clarence G. Michalis, a trustee of the Seaman's Bank for Savings, joined the board. Michalis was prominent in many philanthropic and civic causes in New York City and served on the museum board for 31 years.[4]

Chapman assured the early board members, as he would those who came after them, that they had absolutely no financial responsibility because of their position as trustees. As far as Chapman was concerned, the board existed to fulfill the requirement set out in the bylaws, which directed there be one. It met once a year, read the minutes of the last meeting and the president's and treasurer's reports, elected officers, and agreed to meet again the following year. Until a year before his death in 1959, when he was too ill to be involved, Chapman served as president of the board and made all of the decisions regarding the operation of the museum.

On January 31, 1935, just days after the society was officially incorporated, Chapman wrote Sumner Appleton of his plan to preserve the Tredwell house. It was not so much advice he was after but validation and identification with the most prominent preservationist of the time. Nevertheless, advice was precisely

---

4 Three generations of the Michalis family have served on the museum board: Clarence G. Michalis; his son, Clarence F. Michalis, who became a board member in 1968; and his granddaughter, Helen Michalis Bonebrake, who joined the board in 1992 and at this writing, still serves on the board.

what he got. Appleton wrote:

> As soon as you acquire the property and before you do anything to it, have the place so thoroughly photographed inside and out that the pictures will show practically every square foot of the surface in the condition in which the property comes to you ... This is essential for the reason that the spectator is entitled to know what it is he is looking at. In too many cases he is allowed to go into a house ... and at no time is he warned that a good deal of what he sees was put in by the restorers. After they are safely dead and buried there may be nobody around to tell the true story, which may from then on remain a matter of conjecture.

Unfortunately, Chapman ignored this advice. He may have missed Appleton's point since his intention was to return the house for the most part to its original condition. Who could possibly object to that?

Later he wrote to Appleton, "I have rather fought shy of calling in experts or outside advice, as I have been afraid of confusion and conflicting opinions." Clearly Chapman intended to do it his way—after all, he was paying for it.

2

1935–36: Getting Ready

On April 15, 1935, almost two years after Gertrude Tredwell's death, the deed was transferred from Lillie Nichols to the Historic Landmark Society, Inc., and George Chapman was free to begin the task of transforming the Tredwell home into a historic house museum. Chapman told the writer Helen Erskine what the house looked like when he began his work:

> "Only Dickens could have described how that house looked when we went in to put it in order. The shutters were drawn, dust an inch thick coated rugs, furniture and curtains, and the smell of the past was upon that place."

The first thing he did was to get rid of what he called the "junk." With the help of his wife, he sifted through what the auctioneers had abandoned two years earlier, deciding which objects would be appropriate for display or should be saved and which could be disposed of. He likely found stacks of letters, receipted bills, perhaps diaries. But even had Chapman recognized the historical worth of such items, he would probably have considered them nobody's business and destroyed them. Or it is possible Lillie Nichols or Gertrude Tredwell herself destroyed personal papers for the same reason.

Chapman wanted to create a picture of the house as it would have looked before 1865, while Seabury Tredwell was still alive. Of course he would have to sacrifice a certain degree of authenticity in order to convert the nineteenth-century family home to a twentieth-century public building. The house lacked central heating; a small hot-air furnace heated the rear parlor only, and a few electrical fixtures provided partial illumination in the bedrooms on the second floor.

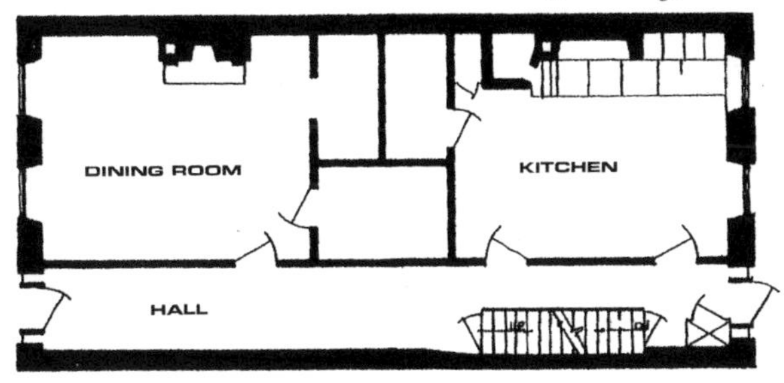

4. Ground Floor

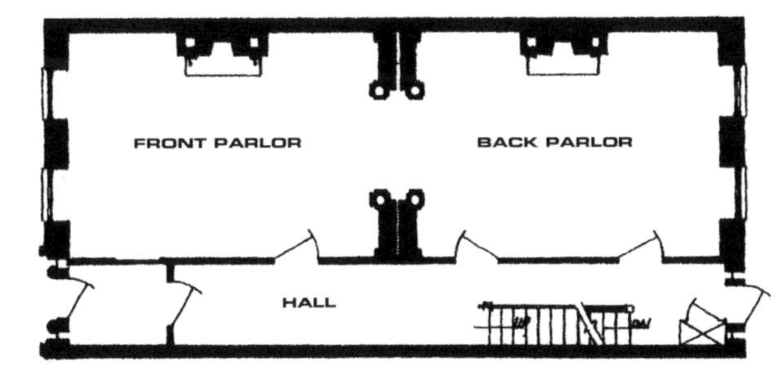

5. Parlor Floor

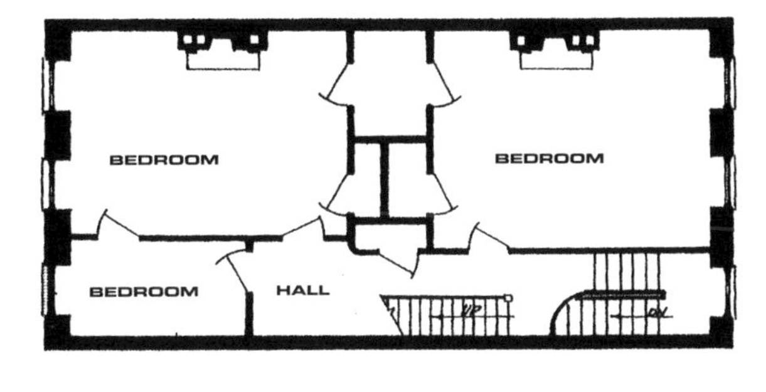

6. Second Floor

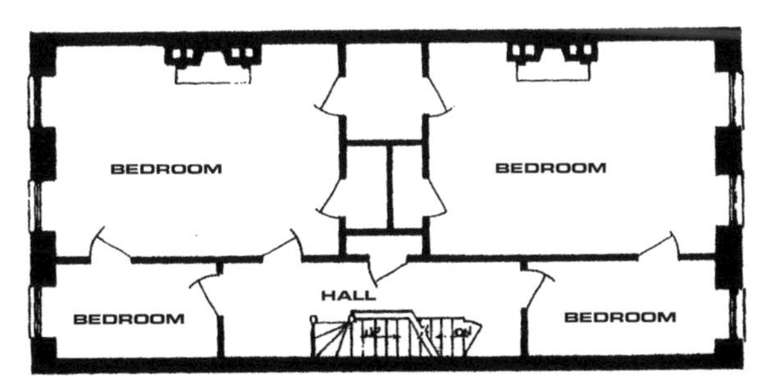

7. Third Floor

Floor Plans

The original floor plan of the Old Merchants House is typical of the 25–foot wide rowhouse of the time. The ground floor comprises the kitchen, a family dining room, pantry, and storage area. Two large rooms of equal size on the parlor floor are separated by sliding pocket doors; on the second and third bedroom floors, they are separated by a pass-through closet.

To address these deficiencies he installed a new coal-fired boiler and steam-heat radiators in all the rooms and fully wired the house for electricity.

The interior layout of the Tredwell home was typical of a Late Federal four-and-a-half-story rowhouse. Two rooms of basically the same size opened off a narrow side hallway on each of the ground, first, second, and third floors. The ground floor, which was a half story below ground, comprised an informal family/dining room on the front and a kitchen in the rear with storage space and a pantry between the rooms. A front and rear parlor on the first floor were separated by sliding pocket doors. A pass-through closet area separated front and rear bedrooms on the second and third floors. The fourth floor, which accommodated the servants, had four dormer rooms and a central work area.

Chapman planned to retain the resident caretaker couple, John and Rita Ansborough, whom Gertrude Tredwell had hired in 1917, to look after the house. The couple slept on the fourth floor in the former servants' quarters but spent their days in the rooms on the ground floor. The front room was their living room; they cooked and dined and bathed in a portable tub in the kitchen, and used a flush toilet that had been installed in a rear-yard shed next to the two-story wooden extension to the house. Now that the kitchen was to be stripped of its twentieth-century appliances and the unsightly rear-yard toilet removed, Chapman would need to somehow provide a small kitchen and bathroom for the caretakers. And because the house was to be open to the public, the City required restrooms for visitors and a fire escape on the rear façade as a second means of egress.

Chapman hired architect Vincent Fox to draw up the plans for the alterations. Fox's plan for the ground floor utilized pantry and storage space between the front room and the kitchen to create a small modern kitchen and bathroom that opened on to the front room for the caretakers and a gentlemen's restroom opening on to the hall for visitors, co-ed public restrooms being unthinkable in 1935 (See figure 8). On the third floor a small hall room was divided, providing space for a ladies' restroom opening onto the hall and a full bath opening onto the front room. The addition of the bath adjoining the front bedroom would make it possible to rent the room in the future (See figure 9).

In keeping with his desire to date the house as early as possible, Chapman elected to remove a sink and toilet that had been tucked under the stairway on the second floor sometime after 1842 when the Croton Reservoir opened and indoor plumbing became possible. He relocated a small tub in the pass-through closet between the bedrooms on the second floor to the caretakers' new bathroom on the ground floor.

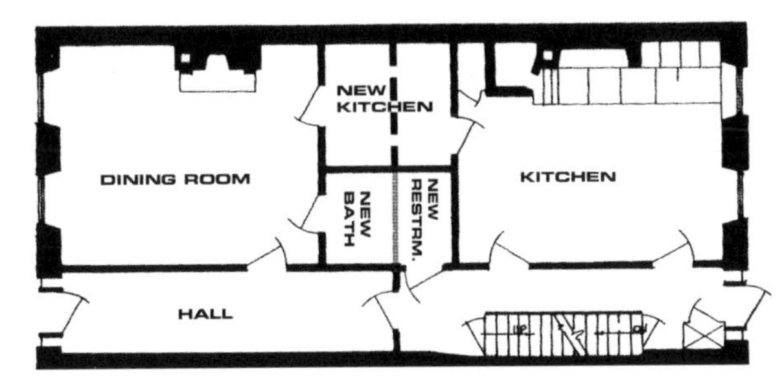

8. Ground Floor Alterations
The dotted line indicates the removal of a partition between the original pantry and a closet, making room for a small modern kitchen. The gray line shows a wall constructed to convert what was a large closet into the caretakers' bathroom and a public restroom.

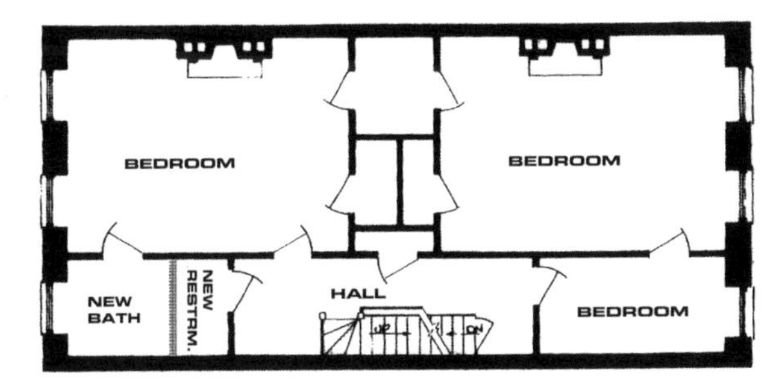

9. Third Floor Alterations
A new partition divided a hall bedroom, converting that space into a public restroom opening into the hall, and a full bath opening into the front bedroom.

In the kitchen, he removed the twentieth-century stove, an icebox, a sink, and two utility tubs located between the north windows that he believed had been installed in the twentieth century.[1] He had a new iron crane made for the fireplace and hung one of the Tredwells' cast-iron pots from it to suggest the open-hearth method of cooking that probably took place in the earliest years of the house.

He converted bedroom closets on the second and third floors into exhibition areas by adding glass-fronted doors, and in the rear bedroom on the third floor, he installed free-standing display cases for the nineteenth-century dresses that had belonged to the Tredwell women. On the parlor floor, in the small room in the rear extension, he installed glass replacements for the wooden doors of the china cabinet where he displayed some of the Tredwells' china.[2]

In addition to these modifications, numerous repairs had to be made. Workmen repaired the window sash and replaced cracked window lights, rehung the shutters, patched the screens, replaced broken panels in the front door, replastered ceilings, replaced rotted beams, rebuilt the fireplace in the kitchen, repaired

1 Recent research revealed these utility tubs dated to the construction of the house in 1832. The tubs have now been reproduced and installed in their original location.

2 This room was originally a porch, enclosed some time in the nineteenth century. It was probably used as a butler's pantry by the Tredwells. The china storage cabinet has since been removed, and the room is now used as a visitors' admission area and gift shop.

the cast-iron stairs leading from the parlor floor to the rear yard, and patched the crack in the marble mantel in the rear parlor. The house underwent a thorough cleaning, including vacuuming the chimneys.

10. KITCHEN, 1936
George Chapman removed all traces of modernity from the original kitchen.

In the early 1870s, the Tredwells had installed a one-person manual lift that went from the ground floor to the third floor for the benefit of Sarah, one of the daughters, who had injured her back while attempting to drive a team of horses. Chapman removed the lift and closed up the floor openings. However, he left the wheel mechanism used to operate the lift in place in the attic above the fourth-floor servants' quarters. He also left in place the dumbwaiter that was installed sometime in the 1870s. Because he thought it marred the look of the back hall on the first floor, he would later have it removed, but for the time being he let it stay. Finally, he had both of the gas chandeliers in the double parlor, as well as various oil, candle, and kerosene lamps, wired for electricity.

On June 19. 1935, Chapman wrote Appleton that he intended to "carefully restore the furniture," to which Appleton replied:

> It is, of course, a difficult thing to lay down a rule just how far to go in the repair and restoration of furniture, but the rule of the Walpole Society seems to be a good one, not to do so much as to take from the "aspect of antiquity," as they call it, of the piece.

Appleton went on to reflect on his own experience of painting the interior of one of the houses that he restored:

> I thank God we didn't take off all the paint. To strip paint and then repaint is to give such a new appearance that any up-to-date apartment house would look just as well as the antique, so why bother to subscribe for the preservation of the antique? Look out you don't lean over too far in the direction of restoration.

While Chapman did not strip the paint, he did thoroughly restore and reupholster the furniture, in spite of Appleton's advice. The last thing he wanted to do was to demonstrate the degrading influence of time on the way of life he wanted to idealize and commemorate. In a letter to Appleton dated November 9, 1935, he explained, "We finally decided to put it all in what we believe to have been its original condition. The fact is, it was so dirty and shabby there seemed to be nothing else to do."

He chose not to display a ca. 1850 rosewood parlor settee and matching table because he thought they were "too late" for the house.[3] He would later place these pieces on consignment with W. & J. Sloane, the upscale rug and furniture store.

Gradually the picture Chapman wanted to create emerged. He restored the paintings; cleaned, relined, and relaid the carpet; and rehung the original red silk parlor curtains. The original lace curtains had deteriorated and were too fragile to hang, so he displayed them behind glass in one of the exhibition closets on the second floor.

---

3 The parlor seating now on display belonged to Amos Trowbridge Dwight, a cotton merchant who lived on West Twenty-Fifth Street. It was a 2008 gift to the museum from the Museum of the City of New York.

11. Front Parlor, 1936

George Chapman deaccessioned some of the parlor furniture because it was of too late a date to be consistent with the period in the history of the house that he wanted to represent. The pictured sofa dates to around 1830.

In the urban homes of the wealthy, it was customary in the nineteenth century for husband and wife to have separate bedrooms. In the Tredwell home, the front and back second-floor bedrooms feature identical large four-poster mahogany beds. Dating to 1835, they were probably purchased when the Tredwells moved into the Fourth Street house. The original bed hangings had disappeared, but in a trunk in the attic Chapman discovered a bolt of burgundy wool damask, which he believed was the same fabric used for the original hangings. There wasn't quite enough fabric for both beds, so after consulting various nineteenth-century design sources, he had bed curtains and swags made for the bed in the second-floor front bedroom and used the remaining fabric to make swags for the bed in the rear bedroom, purchasing similar nineteenth-century fabric to make the bed curtains.

By September 1, 1935, fewer than five months after the transfer of the deed, Chapman was ready to paint the interior walls. Like all historic house museums, the Old Merchants House was to be an edited version of the original, for any amount of restoration, no matter how carefully undertaken, inevitably results in at least one remove from the original. However, Chapman had one inviolate rule that served his purpose well and that he would continue to insist upon as long as he lived. Under no circumstances would he allow anything to be displayed that had not belonged to the Tredwells. Even much of the fabric used in reupholstering the furniture and fashioning the bed hangings had been found in the attic.

12. FRONT BEDROOM, 1936
The original bed hangings had disappeared, but George Chapman found a bolt of burgundy wool damask in the attic and used this fabric to make bed curtains and swags for the bed in the front bedroom.

He was not as scrupulous when it came to the color of the walls. Although the walls had been painted off-white at the time of Gertrude Tredwell's death, he decided to paint the ground floor and the parlor floor what he called "colonial yellow" because he thought that color would show the mahogany furniture and doors to good advantage. On the two bedroom floors, all rooms were to be painted French grey with white trim. The walls of the first-floor vestibule, which had been finished in 1832 to resemble Siena marble, were left as they were. Floors in the bedrooms were left uncarpeted, showing the five-inch tongue-and-groove Southern pine floorboards. Approximately a year after the beginning of reconditioning efforts, he wrote to the New York State Historian that he expected to open the museum to the public in April 1936.

13. Rear Bedroom, 1936
Photography by Henry Fullerton
The swags of the bed valance were fashioned from the wool damask Chapman found in the attic. There was not quite enough of that fabric to make the bed curtains, so a similar fabric was purchased to complete the hangings. Bedroom closet doors were replaced with glass to provide an exhibition area.

But just before the painters were to begin their work, a fire broke out in the small attic space above the fourth-floor servants' quarters. Fortunately, the firemen were able to quickly cover the furniture, which had been moved to the center of the rooms in preparation for the painting, thus preventing water damage. However, the gilded curtain poles from the front parlor, which were stored on the fourth floor, were almost completely destroyed. Enough remained to serve as a model, however, and Chapman had them rebuilt immediately. Two-thirds of the roof had to be rebuilt with a new ridgepole and new beams, and water-damaged ceilings replastered. The cause of the fire was never determined.

By May 1,1936, the house was finally ready for exhibition. As a finishing touch, Chapman placed a hand-lettered card on a table in the hallway, bearing a quotation from William Ralph Inge, Dean of St. Paul's Cathedral in London:

The only promise of a better future for our country is to be looked for from those to whom her past is dear.

To George Chapman, that seemed to say it all. By restoring the home of a successful early merchant—a man of dignity and enterprise—he had provided a place where people could sense the genteel life of a better time and perhaps become better citizens themselves in the process.

Before the museum opened to the public, Chapman entertained approximately 50 guests at a private reception. All afternoon he kept one eye on the door, hoping that Sumner Appleton might attend, but the great man never showed up.

On May 11, 1936, three years after Gertrude Tredwell died and almost 101 years since Seabury Tredwell moved his young family to the fashionable neighborhood of Fourth Street, the Old Merchants House opened as a museum. *The New York Times* took note: "Visitors streamed in all day from the noise and bustle of the street to find the example of a home of a merchant prince of a vanished age."

3

1936–1954: Trying to Survive

The Old Merchants House immediately gained recognition as an important contribution to the understanding of America's historical heritage. In 1935, with the country in the depths of the Great Depression, President Roosevelt had issued an executive order establishing the Works Progress Administration (WPA), a mammoth government agency that would provide relief for millions of desperate Americans who had lost their jobs. Most of the beneficiaries were workers engaged in the construction of roads, bridges, and public buildings. However, one of the divisions of the WPA, the Federal Art Project, found employment for artists. The major undertaking of the project was the Index of American Design, the purpose of which was to compile a collection of watercolors documenting decorative art objects from colonial days through the nineteenth century, including historical costume.

Even before the museum opened to the public, three artists were dispatched to the Old Merchants House to create detailed watercolors of 13 items in the museum's collection including 8 of the 39 dresses that had belonged to the Tredwell women. In addition, a report done by L. Irwin Jones for the Architectural Section of the New York City Unit of the the Index traced the ownership of the Tredwell property from 1640. At the same time, another New Deal agency established by the Department of the Interior began the Historic American Buildings Survey (HABS), a documentation of America's built environment. Out-of-work architects made detailed measured drawings of architectural features of the house and photographed both the interior and exterior.[1]

1 The watercolors of the Index of American Design are housed in the National Gallery of Art. Jones's report, *The Old Merchant's House*, is located in the Library of Congress as are the Historic American Buildings

In August 1936, *The American Collector* featured the new museum, noting that it was "a distinct addition to museum presentations," and the following June, *The Magazine Antiques* called it "one of the most perfectly preserved dwellings of its period."

The operation of the museum went smoothly that first year. In the first week alone, 251 New Yorkers found their way to the unfashionable neighborhood of warehouses and lofts to visit the museum. By the end of year, attendance had reached 1,823. The museum was open seven days a week: 11:00 a.m. to 5:00 p.m., Monday through Saturday, and 1:00 to 5:00 p.m., Sunday and holidays. Admission was 50 cents; schoolchildren were admitted free of charge.

The business of the Historic Landmark Society was run out of Chapman's Fifth Avenue Building office. Beatrice Clark, his personal secretary, played an important role in administering museum operations. She dealt with repairmen, maintained the files, typed the correspondence, and paid the bills. She was also the official secretary-treasurer of the Historic Landmark Society. Each week she put a memo on Chapman's desk stating the amount of the balance in the society's bank account, the total amount of the anticipated bills coming due that week, and the amount of the shortfall. Once Chapman gave the okay, she drew a check on his personal account for his signature and deposited it in the society's account.

Beatrice Clark also kept her eye on the three people who were responsible for the day-to-day operation of the museum: the caretaker couple, Rita and John Ansborough, who continued to live in the house until their retirement in 1937, and Gladys Crutchfield, who served as the museum's curator and guide.

The museum was off to a good start, but in the summer of 1937, just a year after the museum had opened, Chapman, at the age of 67, began to suffer from a troubling and persistent pain in his right arm, which he at first dismissed as a form of neuritis, attributing it to the hot damp weather. But it soon became clear he was experiencing the onset of a crippling arthritis. The disease would eventually affect all of his extremities and permanently confine him to a wheelchair.

As his health deteriorated, he realized that if the Old Merchants House were to survive him, some changes had to be made. With operating expenses running about $4,000 a year, he calculated a $100,000 endowment yielding 4 percent would

---

Survey photographs and drawings.

14. Dress belonging to one of the Tredwell women
National Gallery of Art, Washington, D.C.

A watercolor rendering of a sheer muslin print dress from the early 1860s by WPA artist Roberta Spicer. The artist took the liberty of "repairing" the dress, which was significantly torn with fabric loss at the back, sleeves, and skirt, when she sketched it in 1936. The dress was professionally restored in 2015 and today looks like the watercolor, although in keeping with best conservation practice, the repairs are clearly noticeable and reversible.

put the museum on a self-sustaining basis. In 1938, he began pursuing this goal. It was an extraordinarily ambitious undertaking, considering that this amount would be equivalent to over $1 million today. He acknowledged in letters to colleagues that with the country still feeling the effects of the Depression, it was a terrible time to raise money, but he was going to try.

On April 4, 1938, he wrote to board member Clarence Michalis,

> I am so anxious to try to get the Old Merchants House on a self-supporting basis before the time comes for me to give it up, that I thought at least there would be no harm in sending out an appeal.

And so he composed a four-page appeal for funds, had 10,000 copies printed, along with reply slips and return envelopes, and hired someone to begin sending them out. The Bank of New York would receive the contributions and manage the fund. The first check to come in was for $5. Chapman and Beatrice Clark were ecstatic, believing it was an indication of how the appeal would be received. Even the official at the bank was excited, calling Beatrice on the phone the minute he opened the envelope. But in the end, the results were very disappointing. Over the course of the next year, all 10,000 appeals were mailed; 60 persons contributed $790. After expenses—postage, printing, and clerical remuneration—the museum netted $220. But even though Chapman considered the result "very poor," he continued to take this approach to fundraising, mailing this appeal or others very like it to lists of wealthy persons for the next 20 years. Sometimes he rented lists; sometimes he obtained them from business associates or professional organizations. And each year he culled the social register for new names. There is no evidence that he tried to cultivate the respondents to the endowment appeal, for example, by inviting them to events planned especially for them or encouraging them to form committees to address the museum's needs. Chapman would have considered this sort of thing too troublesome in the best of circumstances. Now, because of his physical condition, it was out of the question.

At the same time he launched the endowment fund drive, he made a tentative effort to see whether he could interest another institution in taking over the museum. He asked his friend Benjamin Wistar Morris, the board member of the Metropolitan Museum of Art, to explore the possibility of the Met assuming responsibility for the Old Merchants House. The answer came quickly. The Metropolitan Museum was not in a position to take over. Morris advised Chapman to look for a private benefactor.

In June 1939, Chapman's wife, Beatrice, died after a long illness. Eight months later, on February 10, 1940, Chapman married his goddaughter, Frances Freeman, in a quiet ceremony in the chapel of St. Thomas Episcopal Church on Fifth Avenue. It was a marriage of convenience on both sides.

With the total contributions to the endowment fund still under $1,000 after two years, it was obvious that the goal of $100,000 was not going to be reached soon, if ever, so Chapman stepped up his search for an institution to take over the museum. Having acquired the Tredwell house and restored it at considerable expense to himself, he would now spend the rest of his life trying to find someone to take it off his hands. In the winter of 1940, the target was John D. Rockefeller, Jr.

With the help of Elinor Robinson, who worked in Mrs. Rockefeller's office, Chapman managed to reach the philanthropist with a letter detailing the historic importance of the house and offering to turn it over to the Rockefeller Foundation, furniture and all. Mrs. Rockefeller passed the letter on to her husband, who notified Chapman in February that he had passed it on to the president of the foundation. But he wrote he was "not hopeful that it will fall within present policies." He explained that because of the large amount of time and effort he was devoting to restoring Colonial Williamsburg,[2] he was unable to participate in such projects. But Chapman, hoping to ward off a negative decision, wrote again on February 29, 1940, pleading for him and Mrs. Rockefeller to visit: "I know that when you see it your civic pride ... will prompt you to lend a hand in its preservation. I feel unable to go on with it and yet am most reluctant to see it abandoned."

In March 1940, a letter arrived from Rockefeller's secretary with the news that Chapman feared but probably anticipated. A restoration project such as the Old Merchants House did not fall within present policies of the foundation.

Still, Chapman did not give up. In his letter of thanks to Elinor Robinson he reiterated his invitation for the Rockefellers to visit. And a year later, in March 1941, they did. They didn't sign the visitors' register, but Harry Lonnberg, who along with his wife, Ellen, had replaced the Ansboroughs as caretakers, recognized

2 Colonial Williamsburg is the largest living history museum in the United States. The restoration of the eighteenth-century capital of colonial Virginia was financed and supported by John D. Rockefeller, Jr. until his death in 1960. The 301-acre historical area includes 88 original buildings and hundreds of other buildings, most of them constructed on their original foundations. Since 1933, Colonial Williamsburg has welcomed more than 100 million visitors.

them—and the initials on their car—and spoke to their chauffeur who "admitted" they were indeed the Rockefellers. According to Lonnberg, Mrs. Rockefeller was "frightfully interested, but Mr. R. was very quiet." And that was the last anyone heard from the Rockefellers.

Chapman was not getting much encouragement from his board at this time, either. Board member Hardinge Scholle wrote on January 10, 1941, "I agree ... that it would be a pity to lose the house, but it is, as we know, almost impossible to finance such ventures nowadays".

When spring came, Chapman confessed to Michalis: "I have not the time or strength to do more than sign checks. I have reached the point where I feel I would not be justified in keeping on any longer."

But still Chapman hung on. Finally, in January 1942, he took a long shot. He wrote to Sumner Appleton, offering to turn the house and its contents over to SPNEA in any way they saw fit. "I hope this will not startle you too much," he began. He argued the connection with New England was "clear and authentic" because Seabury Tredwell's mother descended from New England settlers, including Priscilla Mullins and John Alden of Mayflower fame. It was certainly a stretch, but perhaps he had convinced himself that Appleton's organization might make an exception and expand their operation outside of New England. On January 12,1942, he wrote to Appleton:

> Future generations will rise up and bless you for saving the one example of a fine old New York dwelling and the only one that ever can be saved.... I have reached seventy-two years of age and have been stricken with illness, and unless the right people take hold of this project the whole wonderful example of architecture, etc., will be lost forever.... The man at the helm is knocked out and can no longer handle it.

Appleton obviously did not want to reject Chapman's plea outright. In his reply, he suggested, "The request, while not unheard of, has so far been considered outside the sphere of our legal operation." But then he asked for a statement of income and expenditure for five years, implying that the request would at least come under consideration. But he added, "I'm afraid I can't hold out much hope." Surely, Appleton suggested, there are enough interested wealthy people who could form a group that would be able to make it solvent. He just couldn't understand why, in New York City, such a group could not be found. He even suggested names of people to contact. But the personal glad-handing involved in the cultivation of the wealthy was impossible for Chapman, given his physical condition. In fact, he was in such bad shape he had to hire a fellow attorney, Francis T. P. Plimpton, to furnish the information Appleton wanted and to conduct further correspondence with him. Plimpton's correspondence with Appleton stretched into spring. The SPNEA board would have to approve, and they were not scheduled to convene until April.

In the meantime, although Chapman had never been enthusiastic about government management of historic properties, dismissing such an arrangement as "too political," Plimpton took it upon himself to try to interest Robert Moses, New York City Parks Commissioner, to come look over the house with an eye to the City taking it over. After viewing the house, Moses wrote to Plimpton on March 25, 1942: "Let me be quite frank," he began (Robert Moses being nothing if not frank):

> I thought it a much older house. It dates back to 1830. There are scores of residences which go back as far as that. No doubt many of them lack contemporary furniture, which is good enough in this case, but not particularly distinguished .... It just is not up my alley.

Plimpton thanked him for his "courtesy."

The house so summarily dismissed by Moses was the first site in Manhattan to be designated under the Landmarks Preservation Law of 1965 and today enjoys both exterior and interior landmark status.[3]

Finally, the answer came from SPNEA on April 17, 1942. Appleton was sorry, but the conclusion of the board was that the charter would not allow such an arrangement. Chapman was so disappointed and discouraged that he set August 15 as the date for throwing in the towel. He wrote Appleton on May 13, 1942:

> I will arrange with the Bank to carry on for another quarter to August 15, and if by then I don't have someone I will be disposed to drop it and let the Bank take over, which means all the wonderful old furniture (which belongs to me) would have to be sold and the house go on the scrap heap. It is indeed a dreadful shame to think there is not enough appreciation in NY to preserve this one example of the stately days of yore.

---

3 As of this writing only 117 building interiors have been landmarked. The Merchant's House Museum is one of only six residences to be so designated.

But August 15, 1942, came and went and nothing changed. He had come perilously close to giving up, but he was feeling somewhat better and, although his physical condition continued to deteriorate, for the next 17 years, until his death in 1959, he kept up his efforts to raise an endowment and to find an institution to take over. The endowment fund never quite reached $30,000, and his no-strings offer was turned down not only by The Metropolitan Museum of Art, The Rockefeller Foundation, and SPNEA, but the New-York Historical Society, the Museum of the City of New York, The Cooper Union, the Municipal Art Society, the American Scenic and Historic Preservation Society, and the National Trust for Historic Preservation as well. But he never again seriously considered abandoning the project.

In the spring of 1943, Chapman decided to put the vacant third-floor front bedroom to use by offering it rent-free as partial payment to someone who could do publicity for the house as well as send out the appeal for endowment contributions. He invited applications. The successful candidate was Florence Helm, a graduate of the Spence School, a private girls' school uptown, who maintained a home in Truro, on Cape Cod. She had run a dress shop, a restaurant, and a gallery, but nothing had ever quite worked out for her. She proved to be an enthusiastic booster of the Old Merchants House. No sooner had she settled in than she was planning a party, to take place on June 28. This was to be a real party, with alcoholic punch ("It needn't be too strong," warned Chapman). Florence was determined to inject drama and romance into her promotion of the house, and she understood the need to appeal to the members of New York's moneyed society who, she reasoned, were likely potential donors. Members of the New York Junior League, a young women's volunteer organization, donned nineteenth-century gowns, most of which had belonged to the Tredwell women, and arranged themselves in "tableaux" portraying different periods in the history of the house.

In 1943, no matter what else was happening, the realities of the Second World War were never far from anyone's mind. For almost two years, Americans had been fighting the enemy in Europe and the South Pacific. So it is not surprising Florence Helm decided to feature a historical wartime wedding as part of the party arrangements.

In the rear parlor, a bride and groom and their attendants gathered around the table. The men were dressed in uniforms worn by the US military during the

Mexican War of 1848, complete with shiny gold epaulettes and jangling swords. Clover Dulles, daughter of future CIA Director, Allen Dulles, played the part of the maid of honor. Upstairs in the second-floor bedrooms, tableaux featured a group of young women preparing to leave for church and another group visiting a bedridden convalescent.

This was the first time, but it would not be the last, that someone other than a Tredwell woman donned the nineteenth-century dresses—just for fun. Today, costume curators might well shudder, but in 1943, small historic house museums did not adhere to the professional museum standards that are today taken for granted. The socially prominent guests also sat on the Duncan Phyfe chairs and insouciantly rested their punch cups on every available surface.

Chapman covered the expenses out of his own pocket. The proceeds of $128 were divided between the house and the USO, the United Services Organization, established in 1941 to provide recreational and spiritual facilities for the Armed Forces. All of the New York papers covered the event, which was attended by 125 people, and Florence estimated that 200 persons came to visit the house as a result of the publicity.

In her efforts to bring dramatic interest to the house, Florence Helm sometimes simply made up stories about the Tredwells out of whole cloth—stories about ghosts, love affairs, and romantic balls—which she fed to a willing press. Some of these tales are still circulating today.

Chapman was not entirely comfortable with her creative efforts, and on more than one occasion simply denied her permission to put them into practice. However, one idea seems to have slipped under his radar. In the fall of 1943, Florence arranged for a fashion photo shoot by *Mademoiselle* magazine. Although she convinced Chapman to sign off on it, she was careful not to tell him specifically what was afoot. When the magazine appeared, it featured models in fashionable nightwear—"dream easies"—deployed in the bedrooms, preparing to climb into Seabury's and Eliza's beds.

Beatrice Clark worried about Chapman's reaction, but there is no record of how he took it. The following February, the house received a more dignified treatment in a photo essay in *House and Garden* magazine consisting of four pages of photographs of the rooms and furnishings by the well-known New York architectural photographer Samuel Gottscho.

For over a year, Florence Helm continued to handle publicity, but in the fall of 1944, Chapman had Beatrice Clark inform her he could no longer afford her services and she would have to leave so they could rent her room. But Florence was comfortable where she was; she had a genuine affection for the house and was hurt that Chapman felt he could do without her. She offered to forego her salary and spend one day a week doing publicity in exchange for her rent. Chapman declined the offer. Finally, she convinced him to let her stay on as a rent-paying tenant, and to pay her $5 a week for half a day's work sending out the appeals for contributions to the endowment fund. As a matter of fact, she spent much more than a half-day on the house's business: watching the house on the caretakers' night out (Chapman insisted that the house never be left unoccupied), answering the phone, and replying to letters of inquiry. And when Gladys Crutchfield, the museum's curator and guide, resigned in November, she took over the responsibility of showing the house to visitors. Florence Helm, like so many who would come after her, had fallen in love with the Old Merchants House and was willing to go the extra mile to keep it going.

In December 1944, Lillie Nichols died at the age of 90. Chapman was deeply disappointed to discover from her attorney that she did not leave anything to the museum. He had not expected a financial bequest necessarily, but he thought she might leave something—a piece of furniture, a painting, books, china—something owned by her grandparents, Seabury and Eliza Tredwell, or her mother, Elizabeth Tredwell, which he could put on display.

The Historic Landmark Society had made only interest payments on both of the mortgages since assuming them in 1935. Shortly before Lillie's death, Chapman had persuaded the bank to assign the $14,000 mortgage to him personally in consideration of a payment of $7,000. He had tried for several years to persuade Lillie to cancel her $4,000 second mortgage altogether. Now he began negotiations with her attorney. Finally, in 1947, he was able to purchase the mortgage from Lillie's estate for $300. In 1951, he sold both of the mortgages to the Historic Landmark Society for $8,000, thus ensuring that all indebtedness on the House was discharged.

Chapman had been partially retired for some years, spending the summers at "Airlie," his country home in Mt. Kisco, and frequently vacationing in North Carolina and Vermont. From these remote locations, he communicated with Beatrice Clark by letter and phone and entrusted the running of the museum to her. In 1949, when he officially retired and left his Fifth Avenue Building office for good, Beatrice Clark resigned as secretary-treasurer of the Historic Landmark Society, whereupon Chapman revised his will, revoking a bequest of $20 a month to Beatrice, and Florence Helm took over the position of secretary-treasurer of the society.

Caretakers Harry and Ellen Lonnberg retired at the end of 1953, after 17 years, and on January 15, 1954, Shelly and Phyllis Fox, new caretakers with new ideas, moved into the ground-floor apartment.

34

THIS WAS NEW YORK

ECHOES OF FAMILY MUSIC linger around the square piano in one corner of the front parlor. A hymnbook of 1845 stands open at a favorite selection. The portrait of Seabury Tredwell was painted about 1860 by Henry S. Loup, N.A.

THE CENTURY-OLD HOUSE ON THESE FOUR PAGES IS INTACT, UNCHANGED

When Seabury Tredwell in 1835 moved his family from the hurly-burly of Dey Street up to their new home at 29 East Fourth Street, the neighborhood was fashionable and quiet, the house barely five years old. Mrs. Tredwell could sit in her little tea room in the rear and look over gardens, through magnolia trees, clear up to Fourteenth. From her front windows she could watch other ladies of fashion drive by in high-wheeled carriages and gentlemen on horseback raise tall-crowned beaver hats in salutation.

The clatter of hoofs on the cobblestones was punctuated by the cries of street vendors. "Wud! Wud! Wud!" chanted the wood man. "Shad! Buy any shad!" "Ripe water melyuns!" And in Winter there was the cheerful clang, clang of the muffin man's bell.

Nobody knows what architect designed the house. Some say Minard Lafever, others John McComb—but whoever he was he designed a gracious home. And Seabury Tredwell filled it with the treasures it deserved. Silver and Sheffield plate from England, carpets from France, precious silk and wool damasks, Crown Derby and Limoges—all came in the hulls of his fellow merchants' sailing ships to take their place among Chippendale and choice pieces from the workshops of Duncan *(Continued on page 84)*

THE HORSEHAIR COVERED SOFA was probably made by Duncan Phyfe. Matching doors are of mahogany. The one on the left opens into the hall. The other is a so-called "blind door" put in for balance.

TWO PARLORS with connecting doors rolled

15. ***HOUSE & GARDEN*** **FEATURE (PAGE 34)**
An article about the Old Merchants House appeared in ***House & Garden,*** February 1943, featuring photographs by the noted architectural photographer Samuel Gottscho.

CLOSE-UP OF DINING ROOM. Wedgwood and Crown Derby in the house include wedding china given to Mrs. Tredwell in 1820. Twin windows and pier glass at the end of the front parlor match those in dining room. See below.

Bronze chandeliers for gas were among the first in New York City.

Handsome gilt rosette tie-back set off the crimson draperies

16. *HOUSE & GARDEN* FEATURE (PAGE 35)

4

1954–59: Taking Care

If Florence Helm had been a little more than Chapman bargained for, Shelly Fox was a lot more. The Foxes had been living in Greenwich Village where Shelly was a potter and Phyllis a painter. When he decided to close his studio on Cornelia Street, the chance to be on-site caretakers at the Old Merchants House appeared to be a godsend. Phyllis was to handle the visitors; Shelly, who arrived with a well-stocked toolbox and a thorough knowledge of electricity, plumbing, carpentry, and heating systems, was fully prepared to do what caretakers do.

And that was the problem. It had been over five years since Chapman had been able to negotiate the stairs at the museum, and even when he could, he had paid little attention to maintenance. When circumstances demanded, he hired a contractor to make repairs, but he did not direct Harry Lonnberg to spend money on the kind of preventive maintenance an old building requires. He may not even have been aware of what that kind of maintenance entailed. The Lonnbergs had not revealed the need for repairs to Chapman because they feared, as Ellen Lonnberg confessed to Shelly, they would be accused of "letting things go," which they certainly had.

On March 12, 1954, just two months after the Foxes moved in, Florence Helm died suddenly, after being diagnosed with a case of light jaundice only ten days earlier. After her death, Chapman lost no time in hiring part-time clerical help to work on the endowment appeal and in renting her third-floor bedroom to David Swit, a young man who worked for the Associated Press, and his wife. A baby was born to the couple in December, and for the first time in over 100 years, an infant resided at 29 East Fourth Street.

Eighteen years had passed since the house was opened as a museum, and it was in a dismal state of disrepair. Shelly was quick to enumerate the problems. Broken windows, sagging door frames, loosened hardware, ripped linoleum, cracked plaster, leaking pipes, splintered sash, and chronic dry rot, for starters, not to mention a menagerie of insects and rodents that had taken up residence—"roaches, winged black ants, garter spiders, water bugs, coal fleas, and two generations of mice and pack rats," as Shelly put it in a letter he wrote to Chapman shortly after he and Phyllis moved in. The chimney needed capping, the roof needed work, and the boiler was broken. Shelly repaired the boiler, but he emphasized to Chapman that it was a temporary fix. He also urged him to have the heating system evaluated by the company, noting they were spending an inordinate amount on coal. The Lonnbergs, he learned, were in the habit of scouring the Bowery each night for wood to burn in the fireplace to keep warm.

Over the next year, the Foxes spent $1,100 of their own money in making repairs to their quarters. When Shelly submitted a bill for $39 for supplies he had purchased for repairs to the rest of the house, he proudly noted that if they had had to hire a plumber, welder, electrician, and mason to do what he had done, the total cost would have been over $300. Chapman's response was not encouraging. On March 27, 1954, he wrote:

> I am sure the things you are doing will put the house in better shape and make up for past neglect; however we must keep within our budget and try to get along without any expense except what is absolutely necessary. The alternative I am afraid is that we will have to give up the project entirely.

Still Shelly continued to try to do his job as he saw it, cutting corners where he could, sometimes employing unorthodox techniques to save money, and convincing some suppliers to set up wholesale accounts for the museum because of its nonprofit status. Throughout his tenure as caretaker, his habit was to make a daily round of checkpoints, repairing leaks, patching plaster, making carpentry repairs, fixing what he could, and catching new problems as they arose. And he kept nagging Chapman about the critical need for major repairs. But Chapman did not want to hear about it, privately dismissing Shelly as an alarmist.

In September of the Foxes' first year on the job, however, he had no choice but to pay attention. Hurricane Edna slammed into the east coast, and the storm that hit the city shattered windowpanes in the parlor and ripped metal flashing and slate tiles from the long-neglected roof, causing water to pour through the ceiling into the Swits' room at 3:00 a.m. Shelly was able to replace the windowpanes, but the roof was another matter, requiring the services of a roofing contractor and a rigger. To pay the bill of $700, Chapman found it necessary to dip into the endowment fund, selling ten shares of Standard Oil stock.

Shelly and Phyllis canceled a planned vacation to be on hand for the roofing repairs. And if that wasn't enough, shortly after the storm, the premises were invaded by a species of post beetle that stripped all vegetation within sight before spinning a web that blanketed the exterior walls, Shelly confessed to Chapman that he and Phyllis experienced a "near psychological reverse" when the beetles started coming into the kitchen.

But as dramatic as hurricane-force winds and a Biblical plague of insects might have been, it was the boiler that gave Shelly the most concern. He pointed out to Chapman that visitors constantly remarked on the chilly condition of the house. On the coldest days of winter, the temperature on the parlor floor hovered around a frigid 45 degrees.

But a new boiler was out of the question. Since the beginning, the museum had been a constant drain on Chapman's personal resources. By the fall of 1954, he had made over $50,000 of contributions and advances. The endowment fund, which he had hoped would sustain the Museum after his death, had reached only $27,000. It is hardly any wonder he was loath to spend yet more on the house.

Finally, on October 16, 1954, Shelly, exasperated, went on the record, warning Chapman that the house was approaching a "dangerous state of safety."

> The buses rumbling by, the jolting of the trucks in the garages on either side, as well as the hammering from the heavy 50-ton power presses on this highly industrialized block combined with an era of basic neglect, have subjected the entire structure to strains and vibrations that the architects could not have foreseen.... What should have been constant vigilance on a minor scale was willfully overlooked, with the property value steadily declining and official condemnation looming closer.... I will not be responsible for the potential consequences, nor could I logically be held to account for conditions which were allowed to fester these many years... provided they are allowed to continue unattended.

Still Chapman was unmoved. After enduring another winter in the cold house, Shelly decided to approach Mrs. Chapman. He had concluded Chapman was simply unable to focus on the problems of the museum because of his ill health. A long letter to her in April 1955 laid out two priorities for attention. First was the boiler.

A very technical explanation seems to have worked, for by the summer of 1955, Shelly was negotiating the purchase of a new gas boiler. But of course the boiler was just the most pressing priority; a lot more needed to be done, and much to Chapman's annoyance, Shelly kept up his request for more funds. In September 1956, Chapman finally okayed an expenditure of $295 for the second priority on Shelly's list, repairs to the ground-floor windows and door. He then immediately asked James Wood, secretary of the board, to help him find a couple to replace the Foxes. Wood said he would try, but nothing came of the endeavor.

In 1957, two years after the installation of the boiler, a visitor, noting the decrepit appearance of the house, offered to donate interior painting. Shelly passed on the offer to Chapman but advised him to turn it down. As he pointed out, before any painting could be done, extensive repairs to the peeling plaster in the kitchen would have to be made; doors would have to be re-planed and rehung; and areas around the radiator that had rotted would need to be repaired. "New paint," Shelly wrote, "is the last mile on the long road to restoration." Other more pressing needs, he pointed out, included storm windows for the north side of the house and electrical repairs to augment the inadequate lighting. Chapman declined the painting offer, but Shelly never did convince him to have storm windows installed nor to update the lighting.

Shelly Fox undoubtedly was a thorn in Chapman's side, but he came along at a fortuitous time. Even though he wasn't able to undertake the major projects that he knew were essential for the long-term survival of the house, he pulled it back from the brink—at least for a little while.

Writing to a contributor in 1957, Chapman looked back on over two decades of effort. "We have not done so badly in the 20 years since we took over the Tredwell house," he began. Considering he had done it all practically single-handed, he was entitled to feel a sense of accomplishment. He had rescued a nineteenth-century house that, except for his efforts, would certainly have been destroyed and, through constant infusions of his own money, had kept it going as a museum for over 20 years.

Yet things had not turned out exactly as he had hoped. Early on he had given up the idea of acquiring other properties and being the New York equivalent of SPNEA. He had fallen far short of establishing the $100,000 endowment fund he had envisioned. He had really never attempted to marshal a large enough cadre of reliable supporters who could be depended on to support the museum financially after his death, and his repeated efforts to interest another institution in taking over the museum had failed. And though he still did not completely accept Shelly Fox's judgment, he had to admit the house was in a serious state of deterioration. As Chapman felt his own life drawing to a close, he could not see how the museum could keep going without his financial support. In addition to $37,500 that he had contributed to the society from 1935 to 1956, Chapman had made advances of $14,650 over the same period. Before he died, he persuaded the society to pay him $2,000 in consideration of these sums. Upon payment, it was agreed that the Historic Landmark Society would be released from all debts and claims that Chapman or his assigns might have against it. He also convinced the board to purchase the furniture from him for $5,000, thus transferring the ownership of the chattels to the society. Financial records of the society show these amounts were, in fact, duly paid.

It has been rumored for many years that Chapman's second wife, Frances, as Chapman's heir, somehow managed to force the Historic Landmark Society to pay her $75,000 for the furniture, which belonged to Chapman, thus exhausting the endowment fund. The origin of the rumor is unknown, but its persistence illustrates how tenacious inaccurate information can be, for the story has been around for at least 50 years even though it is false in every particular. The furniture did not belong to Chapman at his death; the endowment fund never came near to totaling $75,000, and the legal agreement Chapman reached with the Historic Landmark Society with respect to the outstanding advances would prevent any claims against his estate on that account. He made no bequest to the museum in his will. Perhaps he felt he had done quite enough already.

In May 1958, his health failing, Chapman finally relinquished direction of the Historic Landmark Society, and Clarence Michalis was elected president of the board. In December, the American Scenic and Historic Preservation Society honored Chapman with a citation for achievement in historic preservation.

> George Chapman, For your unpretentious and sincere advancement of New York's antiquarian culture by rescuing from commercial encroachment and preserving intact the Seabury Tredwell house, generally known as "The Old Merchants House" and for making available to the public this record of a gracious fashion of living now long forgotten save for projects like yours, the American Scenic and Historic Preser-

> vation Society takes pleasure in awarding to you this citation for honorable antiquarian achievement.

George Chapman died at “Airlie” on October 23, 1959. He was 89 years old. Someone less stubborn, less driven, some would say less naïve and arrogant, would probably have given up on the Old Merchants House years earlier. In 1959, it certainly would have seemed to any objective observer that George Chapman’s dream of preserving the home of an early New York City merchant was destined to die with him.

But you just never know.

5

1960–63: Buying Time

By the end of the 1950s, graffiti marred every wall in the neighborhood. Trash littered the tree pits. Drug dealers conducted business on the corners. Drunks slept it off on the steps of the Old Merchants House. Tourists stayed away. The Foxes stuck it out until July 1961, when they decided to move on.

The board closed the house and began looking for a caretaker. Giving up now was not something they were about to do. Clarence Michalis, the board's president, was committed to the cause of historic preservation. Not only had he been on the board of the Old Merchants House almost from the beginning, he had just been appointed to a committee organized by Mayor Robert Wagner to explore ways to preserve historic buildings. Besides, he had promised George Chapman that he would do everything he could to preserve the house. Reminiscing forty-five years after Chapman's death, former board member James Wood remembered that the board did not even consider letting the house go. "We were committed to doing the best we could to preserve George Chapman's dream." But finding a trustworthy caretaker willing to live in the neighborhood would be difficult. The "closed" sign in the window all but invited vandals. Then one day in August 1961, Catherine Roberts, a free-lance writer, walked by the museum, and an improbable chain of events was set in motion that would buy the house some time.

The writer noticed a bronze plaque the New York State Department of Education had installed identifying the house as the home of an early New York City merchant. She was intrigued and, sensing an idea for a possible feature article, she contacted Emmeline Paige, editor of *The Villager,* the Greenwich Village neighborhood newspaper, to see if she knew anything about the house. Emmeline Paige did a little research and when she discovered the board was looking for a live-in caretaker, she had an idea. She and her friend Janet Hutchinson owned a small inn and art gallery in Camden, Maine, which Janet managed during the summers. As it happened, Emmeline had been looking for a place to live, and Janet would probably be glad to escape the Maine winter. What if the board agreed to engage them as temporary caretakers? It would be an adventure for both of them. She broached the idea to Clarence Michalis. The board considered. Something had to be done. If Janet Hutchinson would agree to the arrangement, that would give them more time to find a permanent caretaker, and the presence of occupants would provide some protection for the house.

The apartment on the ground floor designed by George Chapman 26 years earlier had no bedroom, but the third floor had two spacious bedrooms connected by a large pass-through closet, and Chapman had already installed a bathroom adjoining the front room. The renters of the third-floor front bedroom were long gone. All the board would have to do to convert the third floor into a proper one-bedroom apartment was to install a small kitchen in the hall room adjoining the rear bedroom.

The board arranged for work to begin. The costume display cases were removed from the third-floor rear bedroom; the costumes were stowed in the old caretakers' apartment on the ground floor. Workmen installed a stove, a small refrigerator, a sink, and cabinets in the back hall room, and the whole apartment got a new coat of paint. In September, Janet Hutchinson, her teenage son, Jefferson, and Emmeline Paige moved in in time for Jefferson to begin the fall term at Seward Park High School. The teenager would sleep on the fourth floor in one of the servants' bedrooms.

The two women kept the Old Merchants House going while the board discussed what they were going to do next. Years later, remembering the time she lived in the Old Merchants House, Janet Hutchinson recalled that every day from 1:00 to 4:00 p.m., while Emmeline was at work, she showed the house to the few visitors who came to the door. In the mornings and sometimes at night, she polished silver, dusted the old furniture, and swept the fragile carpet with a broom dipped in a bucket of ammonia and water. She also kept busy writing art and theatre reviews for *The Villager*. As theater critic, she made the acquaintance of the Provincetown Players, who performed at the Provincetown Playhouse on MacDougal Street just a few blocks west of the house. After the evening's performance, the actors frequently came by the house, where Vereda Pearson, a popular local entertainer, banged out the latest tunes on the Tredwells' 1840s pianoforte. The actors whooped it up till the wee hours while the house slowly crumbled—quietly falling down around them.

17. THE CARETAKERS, 1961-62
Emmeline Paige (top), Janet Hutchinson, and Janet's son Jefferson examine a trap door between the second-floor bedrooms. The trap door permits access to the mechanism of the sliding parlor doors below.

When spring came, Janet Hutchinson began making preparations to return to Maine, and Emmeline Paige began looking for another place to live. Luckily, by this time, their replacement was waiting in the wings.

Randy Jack was an interior designer who had been intrigued by the Old Merchants House for some time. When he discovered that the board was looking for someone to care for the property in exchange for a rent-free apartment, he contacted Clarence Michalis and applied for the position. In the summer of 1962, as soon as Janet Hutchinson and Emmeline Paige moved out, he moved in. For Randy Jack, the Old Merchants House wasn't just an interesting place to live; it was a cause. He, Randy Jack, intended to save it. And so, with the approval of the board, he tried to interest as many influential people as he could and to raise funds for restoration. At cocktail parties and in informal meetings, he enthusiastically appealed for support. He convinced his numerous contacts in the media to write about the house. Even Hedda Hopper, one of Hollywood's most well-known gossip columnists, mentioned his efforts in her syndicated column, "Hedda Hopper's Hollywood." But while many people expressed interest, funds did not materialize in sufficient amounts to be particularly encouraging. However, in the fall of 1962, Randy came up with an idea that would prove key to the survival of the house.

When Clarence Michalis raised the possibility of finding an organization that would help manage the museum and might be depended upon to supply volunteers, raise funds, and undertake various restoration projects, Randy immediately thought of The Decorators Club.

Founded in 1914 by a small group of women interior designers who were then called "decorators," The Decorators Club is the oldest professional interior design organization in the United States.[1] A group of women started meeting weekly for tea during World War I to discuss their professional interest in decoration and design. While they talked, they decided to aid the war effort by rolling bandages and sewing for the Red Cross. Randy Jack was not a member of The Decorators Club because it was then and continues to be a women's organization. But he was acquainted with Cornelia Van Siclen, a member of the club, and decided to invite her to see the house and ask her if she thought The Decorators Club would want to become involved.

Cornelia Van Siclen, a descendant of Daniel Rapelye, one of Long Island's earliest settlers, was sensitive to the worth of historic properties. She had witnessed the destruction of the Rapelye homestead built in 1650 and razed in 1951 to make way for a sewer development, and she had herself owned a seventeenth-century historic home in upstate New York that had belonged to one of her Dutch ancestors. Those who knew her described her as a "powerhouse" who could get things done and was not easily fazed by difficulty. Just what the Old Merchants House needed.

Her introductory walk-through of the museum was like a tour of Sleeping Beauty's castle, but it was she who was enchanted. With her highly trained designer's eye and her personal background, she recognized the beauty and the historic importance of the property at once. To her, the dinginess of the interior and the tattered condition of the furnishings were not drawbacks but an opportunity. She

1 In honor of the founding members, the club has retained its original name even though the term "decorator" has been replaced by "interior designer" industry-wide. Membership consists of qualified practicing professional women from the fields of interior design, architecture, the decorative arts, academia, and related areas. Throughout the years, the club has promoted high standards of education and ethical practice as envisioned by the founders and has developed innovative and scholarly programs in the study of world culture, design, architecture, and restoration, as well as continuing membership studies of evolving technologies.

was already restoring the rooms in her mind's eye, and as she realized what this could mean to The Decorators Club, she became more and more excited. Randy Jack passed the word of her enthusiasm along to the museum board, and when the invitation to The Decorators Club to act as a sponsoring organization came from Michalis, she couldn't wait to tell her board. In notes for her presentation to the board, she wrote:

> The more I thought about this house the more interested I became with the possibility of it being the home of The Decorators Club. At last we could have a meeting place. Randy was very enthusiastic and he discussed the idea with the trustees and they suggested we could have our office there as well as hold meetings and other affairs. I believed a project like this would give us a reason for being, add great prestige and status to the organization and would have wonderful public relations value.

18. Cornelia Van Siclen
Cornelia Van Siclen, president of The Decorators Club, 1964-65 and 1977-78, and member of the museum board, 1963-78, serving as secretary-treasurer 1970-74. She persuaded The Decorators Club to take on the Old Merchants House as their project in 1963 and was chairperson of the Museum Committee for 15 years,

The board of The Decorators Club agreed it was worth considering and directed her to set up a committee to pursue the idea. On October 30, 1962, the first meeting of the Museum Committee of The Decorators Club was held at the Old Merchants House to determine the advisability of taking on such a big project. The members of the committee were unanimously in favor, but the entire membership would have to vote on it, and so they decided to have a Christmas party to introduce the house to the group. They called it a "Kettle Drum," a nineteenth-century colloquialism for a large holiday party. Girls from the neighborhood Friends School, wearing dresses that had belonged to Gertrude Tredwell and her sisters, sang Christmas carols around a decorated tree.

For over a year the members of the club carefully considered the problems and rewards of undertaking such a large project. The board of the Old Merchants House assured them there would be no legal obligations nor any financial responsibilities beyond those the club itself decided to incur and pay for through their fundraising efforts. Finally, at Cornelia's urging, a poll of the membership was taken. The results were clearly favorable though not unanimous, and on November 6, 1963, The Decorators Club formally accepted the challenge. Randy Jack would continue to live in the house as caretaker, and The Decorators Club would undertake projects to restore the house to its former beauty.

6

1963–65: A New Approach

Decorating, however, would have to wait. The first order of business was a thorough cleaning. The women swept and mopped and vacuumed, cleaned out cupboards, scrubbed and polished furniture and silver and glassware, and had the worn parlor carpet professionally cleaned.

Their approach to the management of the house could not have been more different from that of George Chapman. He had wanted to preserve the house in order to demonstrate the way of life of the early merchant class. He saw the restoration work he did as a means to that end. The members of The Decorators Club, too, looked forward to the preservation of the historic site, but for them it was a professional challenge. They would refinish and recover the furniture, paint the interior, supply appropriate window treatments, and after all their efforts, they would have a restored historic home that would serve as their headquarters. Here they could hold their meetings, sponsor lectures and concerts, entertain at teas, have an office, and store their records. Having met the challenge, their professional prestige, both individually and collectively, would be enhanced. On November 9, 1962, in a letter to Clarence Michalis a week after the original Museum Committee had seen the house, Cornelia Van Siclen wrote, "We want to restore this house to the beautiful house it was originally, so that it will rank with the best of restorations."

Chapman feared confusion and conflicting opinions if he encouraged the advice of outsiders or the involvement of too many people; the members of The Decorators Club figured that confusion and conflicting opinions came with the territory. They did not hesitate to reach out to experts to advise them on their work, and they encouraged members of the club to become involved, not only through their financial support as Friends of the Old Merchants House, but as active members of the Museum Committee. They were accustomed to an organizational structure that relied on a combination of standing and ad hoc committees to get things done. By January 1963, the original seven-member Museum Committee had grown to twenty and had been divided into six subcommittees: Decorating Committee, Fundraising Committee, Research Committee, Activity/Programs Committee, Interiors Committee and World's Fair Committee, which consisted of Cornelia and Randy Jack, who would plan the club's exhibit for the 1964 World's Fair scheduled to open at Flushing Meadows, Queens, in April 1964.

Under Cornelia's firm hand, The Decorators Club managed to keep the museum going during the most troubled years of its history, and eventually in a way they would never have anticipated, The Decorators Club would become instrumental in achieving a restoration that "ranked with the best."

In the years since Chapman's death in 1959, operating expenditures had severely depleted the endowment fund. And although the original understanding with the board of the Old Merchants House was that The Decorators Club would not be responsible for routine maintenance expenses, the museum ran short of funds, and the club soon found themselves paying for repairs and utility bills. Unlike Chapman, they were creative and aggressive in the way they went about raising money, and as soon as they raised it—sometimes even before they raised it—they spent it on whatever needed doing.

Because of their professional affiliation with suppliers to the trade, The Decorators Club was able to secure many in-kind donations for the interior restoration. By January 1964, they had been offered stair and hall carpeting, marble for the vestibule floor, painting for both the interior and exterior of the house, and professional recovering of some of the chairs as well as refinishing of the case pieces. They hoped to have the parlor floor completed for the opening of the World's Fair in the spring of 1964, and they set about getting estimates. However, they soon realized that would not be possible and decided to just lay the hall carpet, install the marble in the vestibule, give the museum a thorough cleaning, and wait until fall for further work on the interior. Time and again the parlor decoration would be put on hold as more pressing needs arose.

In March 1965, the Museum Committee made a detailed estimate of what they thought the restoration would cost. The bottom line was $168,000. However, they earmarked only $37,000 of that amount for structural restoration. The balance was budgeted for interior restoration, including painting, electrical repairs, air conditioning, renovation of the modern kitchen, and new carpeting, draperies, and repairing and upholstering the furniture.

If they could have seen behind the walls, they would have known they were facing problems much more serious than they had imagined. Years earlier, the upper floors of the adjoining buildings had been demolished, exposing portions of the shared party walls of the Old Merchants House—walls that were never meant to be subjected to the elements. Thus, the stage was set for insidious water infiltration. For years, water had been silently seeping through the soft mortar joints in the masonry of the walls. The exposure to moisture and freezing weather caused the brick to decay and the wood inserts for attaching moldings and baseboards to rot. Metal rusted, plaster pulverized, canvas lining peeled away from the wall, and the building settled. Without a major structural restoration—far beyond anything they had contemplated undertaking—the house was inexorably headed for destruction.

Had Cornelia and the members of her Museum Committee understood how seriously the house was degraded and what it would take to save it, they might well have thrown up their hands in despair and hurried back uptown. But they didn't understand, and so they soldiered on.

1961–65: A Significant Recognition

By the early sixties, a small group of New Yorkers had become alarmed at the number of architecturally significant old buildings that were falling to the wrecker's ball. In June 1961, Mayor Robert Wagner, at the urging of architects Geoffrey Platt and Harmon Goldstone of the Municipal Art Society, appointed the Committee for the Preservation of Structures of Historic and Esthetic Importance. It consisted of thirteen members, one of whom was Clarence Michalis. Their job was to come up with recommendations on how New York City's historic buildings could be protected. On November 27, 1961, they suggested that the mayor appoint an advisory Landmarks Preservation Commission to survey New York City's potential landmark buildings and to draft legislation to help preserve them. This advisory commission began its work in 1962.

Meanwhile, architect and preservationist Alan Burnham was at work on an important book titled *New York Landmarks,* which was the culmination of a project begun in 1941 when the Municipal Art Society published a mimeographed list of architecturally noteworthy New York City buildings. The list had undergone several revisions over the years; Burnham's book, published in 1963, represented the latest version. It listed 150 significant New York City buildings that deserved to be saved; the Old Merchants House was one of them. Burnham wrote of the house: "It is probably the only house of its period retaining its original furnishings. For this reason it is unique, and, having survived so long, it must be preserved."[1]

And then in what architecture critic Ada Louise Huxtable, writing in *The New York Times* of October 30, 1963, called a "monumental act of vandalism," workmen

1 Burnham, Alan, ed., *New York Landmarks: A Study & Index of Architecturally Notable Structures in Greater New York* (Middleton: Wesleyan University Press, 1963), 84.

began demolishing McKim, Mead & White's 1910 masterpiece, Pennsylvania Station. The public was outraged. No longer was historic preservation of interest to only a few. As monotonous glass and steel boxes replaced more and more architecturally important buildings, people realized that old buildings gave character, dimension, and beauty to the city. As these buildings were threatened with demolition, public demonstrations became the norm. In a *New York Times Magazine* article of November 23, 1963, titled "Plea to Curb the Bulldozer," the author proposed a selective list of ten New York City buildings that should be saved; the Old Merchants House was one of them.

19. Pennsylvania Station
Library of Congress, Prints and Photographs Division
The demolition of Pennsylvania Station in 1963 galvanized sentiment for the landmarks preservation movement. The beautiful Beaux-Arts structure was first opened to the public on November 27, 1910.

By the spring of 1964, the Mayor's Advisory Commission had finished drafting the landmarks legislation. It called for a permanent Landmarks Preservation Commission of 11 members, which would have the power to designate buildings of historic or aesthetic significance as landmarks. Once designated, the structure could not be demolished until a series of alternatives had been explored, and then only with the permission of the commission. Nor could exterior changes or additions be made without prior review and approval. The commission would have the power to impose criminal sanctions to enforce its decisions.

Now it was up to the mayor. But the legislation sat on his desk throughout the summer of 1964. In September, after another prized New York City landmark building, the Brokaw mansion at Fifth Avenue and Seventy-Ninth Street, was slated for demolition, the proposed legislation moved to the City Council for deliberation. When the Brokaw mansion began to come down in February 1965, New Yorkers winced and howled as stained glass, marble ornamentation, and carved architectural moldings were shattered.

At the same time, a developer who hoped to assemble East Fourth Street lots for commercial use offered to buy the Old Merchants House. This prompted Ada Louise Huxtable to write about its plight in *The New York Times,* February 18, 1965. Michalis was quoted as declaring, "We are in a real crisis. For the last 10 or 15 years, the house has been a problem child. We don't want to sell it but I'm not going to continue this thing any longer. It's been a big headache for a long time. Something's got to happen or it's the end."

The Old Merchants House had survived as a museum for three decades, most of those years by the skin of its teeth. Now it seemed that George Chapman's idea of preserving the 123-year-old building as a museum was just not sustainable after all.

Even schoolchildren knew of the need for landmarks preservation and of the precarious existence of the Old Merchants House. For some time, teachers at the nearby experimental Downtown Community School had included field trips to the museum as part of the social studies curriculum. Aware of the possibility it could be sold and destroyed, they saw an opportunity to teach their students about political activism. On March 6, 1965, a parade of almost 100 Lilliputian protesters marched through the East Village, some carrying placards reading "Don't Throw Beauty Down the Drain," and some playing guitars, singing, "Where have all the landmarks gone? Gone to ruins, most every one. When will they ever learn, when will they ever learn?" After weaving their way through East Village Streets, they gathered at the Old Merchants House where they collected petitions of protest to be sent to the mayor and recited original poems on the steps, for example, "Save the Old Merchants House, please/Or else it will fall on its knees." Cornelia Van Siclen made sure the press and other dignitaries were on hand to greet the children. The wire services picked up the story, and as a result of the publicity, contributions to the museum came from all over the country. By the time the children marched, the landmark legislation was undergoing final deliberation in the City Council. Finally, a month later, on April 6, 1965, it passed unanimously, and Mayor Wagner signed it into law on April 19, 1965.

SAVE THE OLD MERCHANT'S HOUSE

The Old Merchants' House, built in 1832, faces destruction to make way for a garage. Downtown Community School is sponsoring a petition to help save this beautiful house. All New Yorkers should take greater interest in the wanton destruction of old landmarks.

JOIN US IN A MARCH ON SATURDAY, MARCH 6, 1965 TO THE OLD MERCHANTS' HOUSE, AT 29 WEST 4th STREET. THE MARCH WILL START AT 2:00 P.M. AT DOWNTOWN COMMUNITY SCHOOL, 235 EAST 11TH STREET AND PROCEED TO THE OLD MERCHANTS' HOUSE.

20. Children's Flyer

The flyer that publicized the Children's March through the streets of the East Village in support of the landmarking legislation of 1965.

Shortly afterward, the Director of the National Park Service informed Michalis that the Old Merchants House was eligible for registration as a National Historic Landmark. The designation would become effective on January 1, 1966, and be formally granted in a ceremony held at the museum on June 1, 1966. It was an honorary recognition only; the museum would be listed on the National Register of Historic Places and a bronze plaque would be displayed on the front of the building. However, it offered no protection against demolition or inappropriate changes to the building's exterior.

The Landmarks Preservation Commission held its first meeting on September 21, 1965. The Old Merchants House was one of 28 structures to be considered for landmarking. Cornelia was scheduled to speak on behalf of the house. She thought she would be the only one, but as it turned out there were 15 others, including Berry Tracy, curator in charge of the department of American decorative arts of the Metropolitan Museum of Art, as well as representatives from the American Institute of Interior Designers, Municipal Art Society, The Cooper Union, Washington Square Association, County Planning Board, and several New York City historians and interested citizens. Tracy declared the House was:

> . . . a unique, and I stress the word unique, survivor in the City of New York. . . . It remains today inside and out a perfect example of Greek Revival domestic architecture and style. The preservation of the house with its contents is of the utmost importance for both historical and aesthetic reasons.

No one spoke in opposition. The hearing lasted 11 hours—from 10 a.m. until 9 p.m. At the end of the day, the Old Merchants House had been designated a New York City landmark along with 19 other buildings.[2] In the borough of Manhattan, it was the first building to achieve landmark status. This early recognition by the Landmarks Preservation Commission was not only a testament to the historic importance of the house but an indication of just how vulnerable it was, for according to Geoffrey Platt, then chairman of the commission, sites to be considered for landmarking at that first meeting were the most threatened buildings in the city. There was no doubt the Old Merchants House belonged among them.

---

2 Included in the first group of structures to be designated by the Landmarks Preservation Commission was a group of six Greek Revival buildings of Sailors' Snug Harbor and the New Brighton Village Hall, on Staten Island; Kingsland Homestead in Queens; the Pieter Claesen Wyckoff House, the Commandant's House at the Navy Yard, the Naval Hospital at the Navy Yard, and the Prospect Park Boathouse, in Brooklyn; and the Old Merchants House, the Stuyvesant-Fish House, the four surviving Greek Revival rowhouses known as Colonnade Row, the 51 Market Street House, and the U.S. Custom House on Bowling Green, all in Manhattan.

8

1965–68

One Step Forward, Two Steps Back

After three years of trying to make headway with his fundraising efforts, Randy Jack was disillusioned by his lack of progress. Many people had made promises, but they never went beyond the talking stage. It did not appear he was going to be able to save the house after all, and in June 1965, he decided to leave New York City to live in Greece.

He was soon replaced by Sarah Genelli, a young woman in her twenties who with her husband, Tom, took up residence in the third-floor apartment in July 1965. Sarah had a masters degree in fine arts from the University of Hawaii; she was the first person to be affiliated with the museum who had academic training in museum studies. Professionalism in the management of historic house museums was in its infancy, but Sarah had taken courses in the subject and had definite ideas of how things should be done that immediately put her at odds with Cornelia. As far as Sarah was concerned, the museum was being mismanaged by amateurs who were more interested in fundraising and decorating than they were in the educational mission. As the credentialed curator who would be on site and involved in the day-to-day operation of the museum, Sarah presumed she would have considerable say in the way it was run, and she was determined to take a more scholarly approach. This was not the way Cornelia understood Sarah's role. In Cornelia's mind, there should be no doubt who was in charge; she was. While she was not insensitive to the importance of the educational mission, there were more pressing priorities.

But Sarah was energetic and ambitious and determined to put her ideas into practice right away. Visitors to the museum were admitted only on the hour at 1:00, 2:00, 3:00, and 4:00 p.m. Six volunteers from The Decorators Club, trained by Sarah, gave tours. And she was strict about the accuracy of interpretation. She frowned on tales of ghosts and speculation about a trapdoor in the pass-through closet on the second floor being a part of the underground railroad. (Actually the trapdoor simply offered access to the mechanism of the parlor pocket doors below.) Furthermore, Sarah insisted there be "No Touching, No Gum Chewing and No Spike Heels"—proscriptions that Cornelia viewed as so many pedantic affectations.

Sarah also wrote a proposal outlining an ambitious educational program she assumed could be funded by private or corporate donors. It included the scheduling of lecture series and exhibits; publication of articles, post cards, and folio study guides; preparation of slides and a 16 mm film; establishing a reference library; and securing research and curatorial grants for scholars and for herself. Cornelia and the board were willing to support her proposal; Cornelia even met with officials from IBM to appeal for funding. But when that funding was not forthcoming, Sarah felt she was getting nowhere.

Meanwhile, Cornelia and the Museum Committee had turned their attention to restoring the ground floor. They upgraded the electrical wiring, redecorated the front-room reception area, and had the entire floor painted. Once again, they benefited from their contacts in the trade with a gift of an antique chandelier and a reproduction of a nineteenth-century needlepoint carpet. Visitors to the house entered through the door under the stoop and were greeted on the ground floor.

So far, fundraising efforts had been pretty standard fare: lectures, a tour of private apartments and offices that members of the club had been responsible for redecorating, and the Kettle Drum parties. But in the summer of 1965, Cornelia came up with a creative idea that was not only particularly appropriate as a fundraiser for a historic house museum but capitalized on her ability to manage and delegate. The idea was to charter buses that would take participants on a bus tour of New York City historic houses. The tour was called "Revisiting Old New York" and included Van Cortlandt Manor, Dyckman Farmhouse Museum, Morris-Jumel Mansion, Smith's Folly (now called the Mt. Vernon Hotel Museum & Garden), and of course the Old Merchants House, where tea would be served at the end of the tour.

The Steering Committee appointed six committees to cover the bases: Poster and Invitation Committee, Publicity Committee, Narration Committee, Bus Expediters Committee, Ticket Sellers Committee, and Tea Committee. The Publicity Committee evidently did a good job, for the ticket sellers were able to sell 280 tickets at $7. This meant they needed seven buses with a capacity of forty persons each.

The buses, each conspicuously color coded, left from Cornelia's apartment building at 45 Sutton Place South at 10-minute intervals. So for an hour, Sutton Place was transformed into an approximation of the Port Authority Bus Terminal as 280 people boarded buses that pulled out every 10 minutes headed for the Bronx. The plan was that the ten-minute interval would allow the first group to get part way through their tour of the house and out of the way before the next group arrived ... and so on down the line. But of course, more than one bus converged on a house at a time. Thus the need for a bus expediter at each house to keep the "Greens" from debouching before the "Reds" had departed. It was the expediter's unenviable responsibility to see that passengers boarded their buses in a timely fashion. They had to keep up a smart pace, for only thirty minutes was allowed for each visit (ten at the Dyckman Farmhouse). In order to get a running jump on the house tours, a narrator on each bus read from a prepared script about the house coming up next. "The first house we will visit is the Van Cortlandt House in Van Cortlandt Park, built in 1748 by Frederick Van Cortlandt and occupied by his descendants for nearly 150 years.... When the call is given for the green bus, please return to your seats at once."

The tour was deemed a roaring success. The proceeds of $1,400 were promptly spent paying for the downstairs work that had just been completed, and planning immediately began for another tour—"Revisiting Old New York II"—to take place the following June.

The bus tour soon became the signature fundraising event for the Decorators Museum Committee. Once they exhausted sites in the city, they turned to more distant destinations, some of them requiring an overnight stay. Over the years, they traveled up the Hudson Valley to visit historic homes there; to New Haven, Connecticut, for the opening of the Yale Center for British Art; and to Dover, Delaware, to celebrate the state's 200th anniversary. They visited Old Westbury Gardens on Long Island and Governors Island in New York harbor as well as Longwood Gardens and Winterthur Museum in Delaware, historic monuments and sites in Washington, D.C., and the famous cottages of Newport, Rhode Island.

At the Kettle Drum Christmas party in 1965, one of the guests was Joseph Roberto, New York University staff architect. He had grown up in the neighborhood on East Seventh Street. His sister and her girlfriends used to play on the marble steps of the Tredwell house where they were routinely chased off by Gertrude Tredwell herself. He was pleased to see the Old Merchants House was receiving the attention of a group that intended to restore it and to become reacquainted with one of his former students, Ruth Strauss, the treasurer of the club. As they chatted over their punch cups, neither could have imagined what an important role he would soon play in the restoration of the Old Merchants House.

When Sarah Genelli received a scholarship to a six-week institute on museum management to be held at Colonial Williamsburg in the summer of 1966, Cornelia and the board of the Old Merchants House seemed pleased. The board even gave her a stipend for travel, and Cornelia helped arrange for volunteers to give tours of the house in her absence. Sarah's husband, Tom, stayed home to watch the house.

Sarah came back energized—more determined than ever to do things her way, but the priorities of the board and Cornelia had not changed. It was soon obvious she was not going to be able to put her plans into practice, so she resigned, citing in her resignation letter, her "dedication to museum ideals." Her chief complaints were that there was "no budget, no formulated policy for management, and no opportunity to present policy problems and make recommendations to the board." The bottom line was she was unable to "exercise initiative and judgment." Sarah had some sound ideas, but she was years ahead of her time. The Old Merchants House wasn't yet ready for Sarah Genelli.

Her successor was Andre Vernadakis, a young architecture student who had no ideas whatsoever about museum management. He was to sweep and shovel the steps when it snowed, make certain the house was kept locked, give tours on Saturday afternoon, and report to Cornelia. However, because of the demands of his studies, he could never seem to find the time to give the tours. The board had swerved from what they saw as the overly aggressive management of Sarah Genelli to no management at all. The museum was now open only on Sunday afternoons for three hours, when volunteers from The Decorators Club showed up to give tours.

However, a year later, in the fall of 1967, when the Historic Landmark Society board replaced Andre with Jane and Joseph Morasco as caretakers, the house was once again open to the public on a daily basis. Jane assumed the responsibility of conducting tours with the help of volunteers from The Decorators Club. Both she and Joseph were fully committed to maintaining and interpreting the museum to the public.

21. Franco Scalamandre
Photography by Helga Photo Studio
A gift of 100 yards of a reproduction of the original silk fabric used for the parlor curtains was presented to the Old Merchants House by Scalamandre Silks in 1967. Here Franco Scalamandre , founder of the firm, stands before the loom on which the fabric was woven.

If the Genellis had stayed for just a little longer, Sarah would have been thrilled to learn that Lewis Sharp, an architecture student from the University of Delaware, was coming to write his master's thesis on the house. This was exactly the sort of attention she had wanted for the museum. Sharp arrived in the summer of 1967 and submitted his completed thesis the following year.

Of all the donations made to the restoration effort by members of the home furnishings trade, the most exciting by far was the offer by Scalamandre, the legendary textile house, to reproduce the fabric, tassels, and trim of the parlor curtains. The Tredwells had the curtains installed during a redecoration of the house in the 1850s. They had been stored in the attic for many years, reinstalled by Chapman in 1935, and by 1966 they hung in forlorn tatters. Franco Scalamandre himself made a visit to the house to remove one of the panels to use as a pattern for the reproduction of the fabric. By May 1967, 100 yards of the red silk damask had been reproduced, and the elaborate tassels and trim were in production. It was a very generous gift; the value of the silk and the tassels and trim came to $14,950.

The Museum Committee was delighted with the results, but on the advice of Newton Bevin, a respected architect whom they had called in as a consultant, they decided repairs to the outside marble stairs in front, which were tilting ominously, and to the sidewalk iron fence and retaining wall would have to be made before the curtains were fabricated and hung.[1] In addition, they realized that before they could hang the new curtains, the walls and ceiling would of course need painting, but before the west wall could be painted, it would need waterproofing. In addition, the gas chandeliers needed rewiring. Some of the club members had begun to wonder if they were involved in more than they had bargained for. Everywhere one looked there were signs of deterioration and decay. No matter how much money came in, it was just barely enough for necessary repairs and maintenance expenses. Were they not, they wondered, just pouring sand down a rat hole?

To calm the waters, the board of The Decorators Club voted to separate the restoration fund from the club's general funds. Henceforth, the Museum Committee would be responsible for its own activities and expenditures, and those

1 The ornamental iron baskets that had flanked the stairs had deteriorated over the years, and at some point one of them had been stolen. What was left of them had been removed by 1963. The remnants were stored in the cellar.

Decorator Club members who resented the amount of money and attention the house was getting would not be constantly faced with the details and the extent of the problems. The restoration of the Old Merchants House was still a project of The Decorators Club, but it was now a financially autonomous undertaking of the Museum Committee. As head of the committee, Cornelia remained in charge of the restoration.

22. Making the Trim
Photography by Helga Photo Studio
A worker at Scalamandre Silks creating the trim for the parlor curtains.

Of course everybody—especially the Museum Committee—was itching to see the curtains hung. This was the kind of thing they all had signed on for. Their expertise lay in interior decoration—not in making decisions about structural repairs and waterproofing walls. It's a tribute to the good sense of Cornelia and the Museum Committee that in spite of pressure from some of the members of the club, they were willing to grapple with these unfamiliar problems rather than rush ahead with the interior decoration. Franco Scalamandre offered to store the fabric until it was needed, and Cornelia announced at the 1967 Kettle Drum that surely by Christmas of 1968 they would be ready to hang the curtains. But her prediction was wildly optimistic, for as it turned out, the fabric would sit on the shelf for 12 more years.

Throughout the years 1966–68, fundraising efforts were expanded. A letter to members of the home furnishing trade asking for contributions of $100 or more went out in October 1966. Bus tours continued, and club members sold Christmas cards to their friends and business contacts. The Friends of the Old Merchants House had been enlarged and a committee set up under the direction of club member Elisabeth Draper with the goal of raising $200,000 for the restoration. But by June 1967, they had raised only a little over $7,000.

Demands on the treasury were never ending. According to the club's 1967 year-end report, the gutters were "rotting fast," and water cascading down the façade when it rained was causing damage to the front door. In December 1967, the fire department issued a violation for the poor condition of the fire escape and for a hole in the Morascos' third-floor bedroom ceiling. Two days later, while the caretakers slept, the ceiling collapsed, and wet plaster rained down on their heads. Now the Museum Committee faced two more urgent priorities: repairing and repainting the fire escape and relathing and replastering the third-floor bedroom ceiling. And of course, they realized they would have to stop the water infiltration if they were ever going to hang the parlor curtains. So finally, in the fall of 1968, Cornelia and Ruth Strauss, a member of the Museum Committee, called in a waterproofing company to find out what could be done to stop the seepage of water through the west wall, the obvious source of much of the trouble. The contractor convinced them he had the answer: a coat of silicone, he said, would fix things right up. They agreed to have the work done, and Cornelia signed on the dotted line. But for some reason, after the contract was signed and before work began, Cornelia and Ruth began to have nagging doubts about the wisdom of this course of action. They were in over their heads and they knew it.

9

1968–72
Help Is on the Way

It was then Ruth Strauss had the idea of consulting her former professor, Joseph Roberto, the New York University staff architect. He had attended the Kettle Drum party several years earlier and had expressed genuine interest in their restoration efforts. Since he was working on the restoration of his own nineteenth-century home on Sixteenth Street at the time, Ruth knew he would be familiar with the problems of old houses.

When Joe examined the west wall, he immediately saw the waterproofing they were about to undertake was not the answer. The wall was crumbling and would need major masonry repair. As far as he was concerned, any reputable contractor would have realized things were too far gone to benefit from waterproofing. So with the help of attorney Powell Pierpoint, who had recently replaced Clarence Michalis as president of the board upon Michalis's retirement, Joe was able to extricate The Decorators Club from the contract. But masonry work on the west wall would simply have to wait. As serious as the problem was, it was overshadowed by an imminent catastrophe.

To assess the condition of the roof, Joe climbed out on a scaffolding already in place for repair of the dormer windows. He was shocked by what he found. The original gutter on the front of the building, hand hewn from a single 25-foot log in 1832, was attached to a decorative wooden cornice with elaborate hand carved ornaments. Now, 136 years later, the inside of the gutter was so badly deteriorated that the heavy wooden cornice was about to collapse. If it fell, it could seriously injure anyone standing below. His advice was to forget about repairs to the west wall for the time being and replace the gutter and reinstall the cornice. He offered to see to it on a volunteer basis. He would make the drawings for a new gutter, hire a contractor, and make sure the work was done properly, which to Joe Roberto meant perfectly. There was no time to lose.

23. Rotted Hand Hewn Gutter

On January 11, 1969, Joe met with the contractors. Immediately afterward, he began his drawings. At the same time, he assembled an advisory committee of four prominent New York City architects: Newton Bevin, Abraham Geller, Robert Weinberg, and Giorgio Cavaglieri.[1] They would, as Joe put it, "assist and advise in such aspects of the work as will develop from time to time."

Only one departure from the original installation was made. While the gutter was originally hollowed out of one piece of wood, the replacement would be made up of three pieces. This arrangement would be easier to make and more efficient in keeping the water from spilling over. Joe recommended emergency repairs to

1 Cavaglieri was the architect responsible for the restoration of other neighborhood sites. The restoration of the exterior and the redesign of the interior of the Jefferson Market Courthouse for use as a public library in 1961 was one of the first adaptive reuse projects in the United States. Cavaglieri also converted the old Astor Library building to The Public Theater in 1967.

the roof and façade to slow the west wall water damage that by now was accelerating at an alarming pace. The total cost of cornice repair and these waterproofing measures came to a little over $5,000, effectively wiping out The Decorators Club restoration fund.

By the spring of 1969, the cornice was back in place, the Museum Committee was broke, and Joe Roberto was hooked. A new era in the history of the Old Merchants House had begun.

24. Cornice
The replacement of the badly deteriorated original gutter and the reinstallation of the hand-carved cornice in 1969 preceded a total structural restoration of the Old Merchants House.

Joe had discovered that water practically poured through the rotting roof on rainy days. "A full collapse was coming," he would later tell a reporter from *Americana* magazine, "no doubt about it."[2] He knew the Museum Committee did not have the expertise nor the financial resources to manage the kind of restoration that was needed to save the house. In February 1969, soon after he began the drawings for the cornice, he met with the Architects Committee, the Historic Landmark Society Board, and representatives from The Decorators Club. It was clear at the conclusion of the meeting that the cornice restoration was only the beginning of Joe Roberto's involvement with the Old Merchants House. He had committed to rescuing the house from the collapse he saw coming. He knew it would be a major undertaking, but he did not fully realize then just how difficult it would turn out to be. He was 60 years old; he would spend the rest of his life devoted to the restoration and preservation of the Old Merchants House.

The first thing to do was to make a professional assessment of the condition of the house. By April 1969, the Architects Committee was ready with an architectural survey inspection report listing the required work that should not be postponed. Weatherproofing was the most urgent priority. The house needed a new roof, and those parts of the west wall where water was infiltrating and threatening to destroy the ornate plasterwork of the parlor ceilings would have to be rebuilt. Repair of the two-story wooden extension at the rear of the house was also critical. Exposed as it was on three sides to the weather, it was seriously eroded. Out of level and pulling away from the exterior wall, it was literally about to fall off.

Joe estimated that $40,000 to $50,000 would be required for this initial work. But until funding could be found, nothing could be done. The Decorators Club could be helpful in organizing a fund drive, but to raise this kind of money, it would be necessary to contact foundations and state and federal agencies—organizations that would be able to provide thousands or even tens of thousands of dollars at a time.

For the next year and a half, Joe spread word of the danger the house was in, trying to find sources of support. He and his wife, Carol, spent hours writing letters to government agencies and private foundations. He and Powell Pierpoint approached numerous persons in the preservation community who offered suggestions and did what they could, but still the necessary funding eluded them. In Joe's words, "It all seemed so hopeless." And then finally—the turning point came.

---

2 Roger M. Williams, "Landmark on East Fourth Street," *Americana* 9 (September/October 1981): 66–72.

Among the recipients of the Robertos' letters was Ada Louise Huxtable, architecture critic for *The New York Times.* In February 1965, she had written sympathetically about the house when it came close to being acquired by a developer who wanted to tear it down. Now, six years later, Joe thought she might be willing to publicize the plight of the Old Merchants House once again. When his letter arrived, it just so happened she was working on a piece about historic preservation. Titled "A Funny Roll of the Dice," the long article concerned the efforts of Grand Rapids, Michigan, and Boston, Massachusetts, to save their city halls. But determined readers who persevered to the final three paragraphs read about the Old Merchants House in what amounted to a postscript:

> And at 29 East Fourth Street, the Old Merchants House may not get through the winter. This gem of a house museum, the Tredwell mansion that has survived intact and unchanged, but woefully deteriorated, from its days of Greek Revival glory in the 1830s to the death of the last Tredwell lady in 1933, can no longer keep the weather out. Its fine plaster ceilings could disappear with one good storm.
>
> The desperate efforts of the Historic Landmark Society, which owns it, The Decorators Club, which maintains it, and the New York University Architect, Joseph Roberto, who supervised the rescue of the cornice, have not been enough. The need is for $40,000 for immediate structural work.
>
> Have you heard about the New York communications gap? All those rich people, foundations, and corporations sitting around in committee meetings wondering how to give their money away while the best things in the City are going under for the last time. Anyone for some nice civic-spirited Christmas gifts? Seasons greetings one and all.[3]

Those were the words that finally shook the money tree and would make a thorough restoration of the house possible. The result, according to Joe, was "electrifying." The first to respond to the *The New York Times* article was Joan Dunlop, Assistant Director of the Fund for the City of New York, who called and expressed her interest. Eventually the fund would come up with a grant of $5,000, but even more important was her assistance in putting Joe in touch with people whose

3 Ada Louise Huxtable, "A Funny Roll of the Dice," *The New York Times,* December 17, 1970.

involvement would finally lead to major federal funding that, when supplemented by matching funds, would provide the momentum to sustain the restoration through to the end.

One of the persons Joan Dunlop contacted on Joe's behalf was Mark Lawton, Director of the New York State Historic Trust in Albany. She was hoping that it might be possible for the Historic Landmark Society to deed the house to the trust, but to qualify, the house would have to be owned by the City.[4] However, Lawton noted a limited amount of 50 percent matching federal funds for emergency repairs, administered by the New York State Historic Trust, was available to administrators of National Register properties for fiscal year 1971-72. Lawton advised Joan Dunlop to have Joe send a cost estimate for the work necessary to prevent further deterioration. So Joe sent Lawton an itemization totaling $60,000, and on March 12, 1971, he and Powell Pierpoint, met in Joan Dunlop's office with Harmon Goldstone and Geoffrey Platt, the chairman and vice chairman of the Landmarks Preservation Commission, to discuss funding possibilities. By the time of this meeting, Joe had been elected to the museum board, the house had made it through the winter after all, and now it seemed there was a prospect for real money.

In May 1971, the New York State Historic Trust made application to the Department of the Interior for a matching grant in the amount of $30,000, one-half of Joe's estimate of the total that would be necessary for emergency work needed to save the house from destruction. If approved by the federal authorities and matched with a like amount, the Old Merchants House would at last have a new lease on life.

Major progress towards raising these matching funds came from the Mabel Brady Garvan Foundation. Anthony Garvan, like Joan Dunlop, had been motivated by *The New York Times* article to consider the Old Merchants House for funding. Garvan offered $20,000 to be disbursed through the National Trust for Historic Preservation. With the $5,000 offered by the Fund for the City of New York and the Garvan grant, Joe was well on his way to reaching the goal of $30,000 in matching funds.

---

4 In 1967, the Historic Landmark Society Board had tried to make an arrangement whereby the house would be deeded to the City, and the Museum of the City of New York would operate it. The contract was actually drawn up, but the Museum of the City of New York did not go along and nothing came of the discussion.

25. Ada Louise Huxtable and Joe Roberto

Photography by Marilee Reiner

Ada Louise Huxtable, architecture critic of *The New York Times,* and Joe Roberto at the first party to be held in the restored double parlor, a benefit for the International Human Assistance Program in December 1980.

In April 1971, the Greenwich Village Homeowners Association raised $1,500 through an old-fashioned fair, the brainchild of Lee Roberts, a member of the association and an architectural researcher whom Joe would later enlist to research the history of the Tredwell family and the building of the house. Villagers turned out in force to buy homemade jellies, plants, hand-crafted tea cosies, and each other's white elephants to help their Village neighbor. A donation of $2,000 from Mr. and Mrs. Robert Wilson brought the total to $28,500. The Decorators Club turned over the proceeds from two Hudson River sailing parties, and the Friends of the Old Merchants House provided the rest of the matching funds.

In August 1971, Powell Pierpoint learned that the federal authorities had approved the application and the Old Merchants House would receive the $30,000 grant from the Department of the Interior. He immediately appointed Joe restoration architect, with the authority to enter into contracts and to pay the bills associated with the restoration.

The initial grant was intended to cover work needed to forestall further deterioration including rebuilding the extension at the rear of the house, restoring the exterior walls, replacing the roof, repairing gutters, rebuilding the chimneys, and repairing windows, lintels, and shutters, as well as some basic electrical work. As soon as Joe got the good news, he began preparing specifications for the rebuilding of the extension.

One day in August, as Joe was making his plans, the Old Merchants House had a surprise visit from the celebrated Broadway actress Helen Hayes and her friend, Anita Loos, the playwright. The actress had just retired from the stage, and the two women had decided to spend over a year exploring the city, visiting sites unfamiliar to them, including the Old Merchants House. They were accompanied on that day by Tom Ellis, a young actor, and Betty Kaufman, Director of Nursing at Bellevue Hospital. When Betty realized where they were going, she observed that the neighborhood was where they picked up most of their stabbing cases. A sign on the door informed them that the house was closed for the month of August, but not to be deterred, Tom persistently rang the bell. When Deborah Davis, who with her husband had replaced the Morascos as caretakers, stuck her head out the upstairs window, Tom convinced her to open up for the famous visitor.

Deborah apologized for the shabbiness of the place, explaining, "There just isn't enough money to do all the things we should do."

"Just the same," the actress replied, "the old place is still here with its original furniture, drapes, and chandeliers. No amount of rust and wear can keep their beauty from shining through." As she left, she remarked to her companions, "With a little financial help they could give New York City back one of its treasures in pristine order."[5] It would take over nine years and a lot more than a "little" financial help, but she was right: the treasure was on its way back. Helen Hayes helped, leaving a donation of $10.

By the beginning of 1972, Joe had completed his specifications for the work covered by the $30,000 grant; matching funds were in hand, and the way was cleared for work to begin on the structural restoration of the exterior of the house. Joe, however, did not intend to stop there. He envisioned a four-phase total restoration of the Old Merchants House. The first two phases would entail the restoration of the exterior, which was about to get underway and which the $30,000 grant was intended to cover. Phase Three would include interior structural work necessary to prepare the house for Phase Four, which would include replastering and painting, and restoring and reinstalling the interior furnishings. Joe estimated that the cost of the entire restoration would come to $250,000 and hoped that in the future, matching federal grants could be secured for Phases Three and Four.

On January 18, 1972, before work began on Phase One, Joe met in Cornelia Van Siclen's Sutton Place apartment with members of the Museum Committee of The Decorators Club. Giorgio Cavaglieri, representing the Architects Committee, was also there, as was Joe's wife, Carol, who was an interior designer, although she was not a member of The Decorators Club. She had strongly objected to his taking on the restoration of the Old Merchants House for the simple reason that she didn't think it could be done. It was not an unreasonable opinion.

In his remarks at this meeting, Joe attempted to spell out his motivation for undertaking such a difficult project. Why, he wondered aloud, were they all dedicated to "keeping our little building from falling apart?" The idea, he declared, was "startling." "What is the justification," he asked—as no doubt Carol had asked him many times over—"in our expending the effort, the time, the many skills that demand so much of us?"

He noted that it had become commonplace to abandon old buildings and concluded that to go along with the trend and just let the Old Merchants House go "would be for us to lose our heritage, our ties with the past, to break with those

5 Helen Hayes and Anita Loos, *Twice Over Lightly* (New York: Harcourt Brace Jovanovich, Inc., 1972), 304–06.

who succeeded in creating a patrimony for us. They left us a bit of history."

These were not just fine-sounding words. As a major in the U.S. Army Air Force during World War II, Joe Roberto had worked with a team that assessed the effectiveness of the V-2 rocket bomb attacks on London. After the war, he was sent to evaluate the effectiveness of Allied bombing raids all over Europe. Seeing the destruction of centuries-old classical buildings up close, kicking around in the rubble, and documenting the devastation awakened in him an appreciation for classical architecture. He said it was this wartime experience of seeing what Europe had lost that helped him to understand how important old buildings were to the life of a city.

"Now," he went on, "we are making our own history.... We are going to overcome the deterioration of time. We will rebuild, renew, and restore to its former beauty the entire building."

Overcoming the deterioration of time is not normally an option for mere mortals. But this presumptuous activity is what historic preservation is all about. Joe realized that what he was embarking upon would require all the energy, the creativity, and the skills he possessed. It was precisely because the endeavor would demand so much of him that he was determined to undertake it. The unspoken question was, "Was he up to the challenge?"

"If I can succeed in this work," he would later write to an official at the Department of the Interior, "I will have helped to build up the area I lived in as a boy."

George Chapman saw the deterioration of the neighborhood as a sign of the regrettable decline of a social class. Joe Roberto took it more personally.

10

1972
Conflicting Visions

In December 1971, when major funding for the restoration had become a possibility and Joe began planning a four-phase restoration that would cost a quarter of a million dollars, some of the members of The Decorators Club realized that he was about to assume a measure of responsibility that would threaten their authority as stewards of the house.

On February 4, 1972, as the restoration was about to begin, The Decorators Club held a meeting. Anne Winkler, seeing the handwriting on the wall, suggested mention be made in the minutes of the role The Decorators Club had played in saving the house and making it possible for this new phase of reconstruction to begin. Joe and Carol Roberto were present at that meeting and listened as the committee tackled the agenda: Cornelia reported on bids she had received for painting, upholstery, and installing a concealed light in the outer vestibule. They discussed the possibility of upholstering the Tredwell window seats for use in the parlor and whether or not the dumbwaiter should be replicated. Cornelia announced research was underway to discover appropriate paint colors or wallpaper for the hall as well as hall and stair carpeting. They considered an upgrade of the modern kitchen, which they hoped might be underwritten by a donor. Representatives from a carpet manufacturer presented a proposal for reproducing an exact duplicate of the parlor carpet, which would require 200 linear yards of carpeting and cost a little over $6,000.

Before any of their ambitious plans could be realized, however, a vast amount of structural work would have to be undertaken. When it came time for Joe to report on the grant money they had received so far, he reminded them that priority of work had to be given to major restoration work such as roofing, rebuilding the chimneys, deteriorated brick walls, and stonework.

Joe may have felt the members of the committee were getting ahead of themselves, but neither he nor they could have foreseen how difficult a road lay ahead or just how strained their relationship would become before the goal of restoring the Old Merchants House was accomplished.

Unfortunately, the organizational chart the Historic Landmark Society Board drew up in 1971 failed to lay out a clear line of authority for the direction of the restoration. Joe and Cornelia were designated co-chairpersons of the Restoration Committee under the direction of Powell Pierpoint, president of the board. Joe was to be responsible for the exterior and interior structural restoration of the building, and Cornelia the restoration of the room furnishings. However, the administration of the government grants would require that one individual be responsible for the authorization of the expenditure of all funds and have the authority to make final decisions. Powell Pierpoint was obviously not in a position to serve in this capacity. Joe Roberto as the restoration architect was clearly the person in charge, and everyone involved in the restoration would necessarily be responsible to him, no matter what the organizational chart said. Nonetheless, Cornelia had good reason to believe that she was in complete charge of the restoration of the furnishings with the authority to make unilateral decisions and to commit the funds raised by the Friends of the Old Merchants House and her Museum Committee.

No mission statement for the museum existed. There had been no agreement on why they were all saving the house or specifically how it would operate after they had saved it. The board of the Old Merchants House would, of course, be responsible for determining how the house would be used once the restoration was complete. In March 1975, a committee was formed to come up with a policy of house utilization, but they did not reach a conclusion.

As it was, Cornelia and her committee and Joe Roberto were privately pursuing very different goals. The Decorators Club envisioned a "working house" where they could hold their meetings and entertain in a beautifully restored setting that would showcase their professional skills as interior designers. Utilizing historic properties for such a purpose is a legitimate option in the field of historic preservation and is not an uncommon practice. In fact, Powell Pierpoint had encouraged The Decorators Club in such a plan when they first considered taking on the Old Merchants House as their project in 1962. There was, however, a good reason the Old Merchants House should not serve The Decorators Club as a "working house": it was unusual among historic houses in that it still contained original furnishings, many dating to before the Civil War. As an intact home, it represented a unique resource for the study of domestic life in the nineteenth century. In order to fulfill that purpose it would logically need to become a more or less static museum, not a "working house," where historic furnishings and decorative objects would be subject to wear and tear, and loss.

The Decorators Club was sensitive to the importance of the house as an historic document and intended to restore the interior appropriately. They had secured the significant donation from Scalamandre of 100 yards of a reproduction of the original red silk damask curtain fabric, and they were planning for an accurate reproduction of the parlor carpet. While they often accepted donations of valuable antiques, their policy was to raise funds by selling these donations at auction. For the most part, they followed George Chapman's policy of exhibiting only Tredwell furnishings. However, their plans for a "working house," if realized, would inevitably lead to the degradation of the furniture.

Enthusiasm for the Colonial Revival, which was typical for the time, led the Museum Committee to interpret the original kitchen inaccurately, but except for installing marble flooring in the vestibule, so far they had done nothing that could not easily be reversed. Nevertheless, they had been considering installing a sink with running water in the rear extension, an amenity that of course was not historically accurate. They intended to make the house available to the viewing public, including schoolchildren, but their primary goal from the beginning was to restore the house as their clubhouse, and they were not averse to making some improvements to facilitate that objective.

For Joe Roberto, the restoration was an opportunity to demonstrate his expertise as a restoration architect, capable of restoring an 1832 house as nearly as humanly possible to its original condition. He went to great lengths to reuse original material where he could. He saw the house as an historic museum that would be available for scholarly study and serve to increase the public's awareness and understanding of domestic life in the nineteenth century.

While Joe realized none of the members of The Decorators Club fully grasped what they had gotten into when they took on the responsibility for the Old Merchants House, he did not foresee the extent of the deterioration of the building and the difficulties he would encounter in putting it back together. In March 1972, with the restoration barely under way, he wrote Newton Bevin, a member of the architects committee: "Problems are overwhelming at times. What has to be done takes more than one person." But it was Joe who wrote the grant proposals, set the agenda, planned the budgets, drew up the specifications, hired the craftsmen, supervised the work, paid the contractors, justified expenditures, and figured out how to solve the unique problems that cropped up over and over as the work progressed. The restoration placed an almost intolerable demand on his time and energy. Not the least of his problems were the onerous procedural demands of government grants administered by the State, with which he was initially unfamiliar. He had neither the time nor the patience to consult with the members of The Decorators Club during the structural restoration. Since they would not be involved except for fundraising until his work was done, he did not always communicate the details of the technical problems he faced or explain the bureaucratic delays that sometimes slowed the work to a crawl. As a result, during the years that the house was closed for structural restoration, The Decorators Club didn't always have a clear idea of what was going on or why Joe's work was taking so long.

To make matters worse, Cornelia and Joe were polar opposites. He was obsessively meticulous in everything he did. Cornelia, on the other hand, had a more *laissez faire* approach to some of the aspects of her work. She often delayed paying the electric bill for several months at a time, a habit Joe would discover when one day a workman showed up to shut off the electricity for nonpayment. And she was known to come to board meetings with an incomplete accounting of what her committee was up to.

It is no wonder that the relationship between the Decorators Club Museum Committee and the restoration architect became more and more contentious as the restoration dragged on.

26. Kitchen as Interpreted by The Decorators Club, 1965
The rocker, the mahogany table and chairs, the book with its ribbon marker, and the decorative plates would not have been found in a 19th century working kitchen, though these objects did belong to the Tredwell family. Their placement in the kitchen reflects enthusiasm for the Colonial Revival style of decoration in the 1960s.

11

1972–74
Getting Down to Work

The grant process was a comic portrayal of bureaucracy in action. In a Byzantine procedure, paperwork shuttled back and forth between Joe, the New York State Historic Trust (later replaced by the State Parks and Recreation Department), and the Department of the Interior. This went on for months—sometimes in excess of a year—before a project was approved.

First, New York State bureaucrats had to approve a preliminary application. They then submitted an "apportionment warrant" to the National Register Office, the government bureau that authorized the federal grants available to sites designated as national landmarks. This document enumerated all of the projects that State officials deemed eligible for funding and the necessary funding for each. The National Register Office then determined the total amount of money they would allot each State. The allotment was always less than the State requested, so the State bureaucrats then had to decide which of the grant projects they would recommend for funding. After they requested and approved detailed plans and specifications from the successful applicants, they submitted a "Project Initiating Letter" to Washington. Federal officials reviewed and approved the final application, and work could then begin. And because the Old Merchants House was a New York City landmark, Joe was required to submit specifications for exterior work to the New York City Landmarks Preservation Commission for approval as well.

All projects had to be completed within a three-year grant period, and the State provided strict guidelines of the types of allowable costs. Expenditures had to be documented with copies of the front and back of cancelled checks. It was not uncommon for the State to disallow some expenditures or to request further documentation. Upon approval of the request for reimbursement, the State then sent the request to Washington, and a check for 50 percent of the total was sent to the State to be returned to the applicant.

Phase One restoration work on the exterior of the Old Merchants House began in February 1972. The expectation was that the entire restoration would be completed within four years and the reopening of the museum to the public would coincide with the Bicentennial celebration of 1976.

Realizing that more funding would be required to complete the work planned for Phases One and Two, Joe immediately applied to the State for a supplemental grant in the amount of $20,000. The State reduced this amount, to $12,000 before passing the application on to the federal authorities for approval.

Joe began the restoration by rebuilding the extension at the rear of the house. He removed windows, siding, and doors, first carefully numbering and photographing all pieces and recording their placement on a diagram. What he could not restore to its original condition he replaced with new material conforming to the appearance of the original. He added insulation and raised and leveled floor beams after removing deteriorated supports. Finally, he repaired the cast-iron stairs leading from the parlor floor extension to the rear yard.

Mortar joints and the brick masonry of the exposed party walls on both the east and west sides of the house had deteriorated badly.[1] To restore them, Joe salvaged 2,500 bricks from a recently demolished 1830s house in the neighborhood. The bricks were repointed with mortar duplicating older mortars, and a final cementitious coating of waterproofing applied.

Joe made a decision to do some work that he justified on aesthetic grounds. He applied to the Landmarks Preservation Commission for permission to remove the tall parapet walls on both the east and west thus forming a new line following the pitch of the roof. These walls were a source of water infiltration, and Joe thought they were also an eyesore. Rather than rebuilding them, he wanted to remove them altogether. Part of the original construction of the party wall, they were intended to prevent the spread of fire from one roof to another. Joe consid-

1 The brick originally used in the construction of the party walls was known as "Salmon" brick. These bricks, because they are located at the edges of the brick-firing kiln, are less vitrified and much softer than other brick. As a result, they are less expensive and are used generally only for concealed locations like interior walls or the party walls shared by adjoining buildings. When exposed to weather, they can deteriorate rapidly.

ered this undertaking so important that he hired a model maker to construct a scale model with removable parapets to show what the house would look like when the walls were removed.

27. Model of the Old Merchants House
The model of the house Joe Roberto presented to the Landmarks Commission when he requested permission to remove the parapet walls. The removable parapet of the model shows how the house would look without the walls, the original purpose of which was to prevent the spread of fire to an adjacent building.

On May 11, 1972, he presented the model at a Landmarks Preservation Commission hearing, hoping it would convince the commissioners to grant the permission he sought. They agreed to allow him to reduce the height of the parapets, but stipulated they extend 12 inches above the slope of the roof to conform to the minimum standards existing in 1832.

In August 1972, roofers began removing the existing slate shingles on the steeply pitched roof, salvaging those found to be in good condition. They removed and replaced the rotted and defective roofing boards and sheathing and dismantled and rebuilt the two chimneys using whatever bricks they were able to salvage from the old chimneys.

Joe ordered new black Bangor slate from a quarry in Pennsylvania, which he believed had furnished the original slate when the house was built. The order was delayed due to flooding of the quarry, finally arriving in October 1972.

No sooner had work begun on the roof than the Old Merchants House received notice from the Buildings Department that a hazardous condition violation had been placed on the property because of a bulge in the brick wall underneath the front windows of the parlor. Joe knew the bulge was there and had scheduled its correction for a later phase of the restoration, but now it had to be addressed immediately. He discovered the builder had neglected to tie the façade brick to the backup wall of brick and rubble. He had the wall rebuilt below the level of the parlor windows, and also had ties secured to the floor joists installed at the first- and second-floor levels. The front façade wall was now more secure than it had ever been. The City was satisfied, and the violation was removed in December 1972.

Nineteenth-century methods of construction also caused foundation walls to fail on the north and south sides of the house. When Joe discovered open joints where foundation stones had loosened and dropped out, he excavated the area and found the drains were not connected to the pipe system but to dry wells that had been dug close to the foundation walls. It was seepage from these wells that had caused the foundation to fail. The solution was to install new storm drainlines to drain the rainwater from the eaves.

Other projects undertaken during the exterior restoration included sanding and repainting the front shutters; resetting the front steps; patching the marble on the façade; repairing and repainting the fire escape; and replacing all cracked windowpanes. In addition, preliminary electrical work was done, and restoration of the exterior ornamental ironwork begun. In the spring of 1973, the $12,000 supplemental grant Joe had applied for in February 1972 was approved by federal authorities.

By the summer of 1973, with the completion of the weatherproofing work scheduled for Phases One and Two plus the unplanned repair of the foundation; replacement of the dry wells with storm drainlines; expedited removal of the bulge; and rebuilding of the exterior south wall, Joe believed he had successfully completed the crucial work of making the house watertight. However, on August 2, 1973, in

28. Roofing restoration in process, 1972
Photography by John Bayley

checking the house after a heavy rain, he made a startling discovery. Water was seeping through the west wall adjacent to the fireplace in the kitchen. Attempts to trace the leakage to the obvious suspects—the roof, the eaves, the roof of the garage next door—were unsuccessful. It was not until he had the workmen remove plaster and bricks and finally flooring from the kitchen that he determined the source of the water was underground seepage coming from the west. Also, as a result of years of water infiltration, the ends of the beams supporting the kitchen floor were seriously deteriorated. The kitchen floor was headed for collapse. To solve the problem, Joe turned to the ancients. Using a method similar to that employed by the Romans in heating the baths in the Caladarium, he dug out soil and compacted clay between each of the round log floor beams and replaced it with gravel, placing two rows of perforated brick within each gravel bed. Joe believed the air channels thus created would relieve undergrade moisture accumulation.

Finally, in April 1974, the system was in place. Joe was confident it was working, but to make absolutely sure it would disperse whatever moisture might result from heavy rain or snow, he would leave the floor uncovered for another year, periodically checking the moisture level. Before resetting the hearth and replacing the floor boards, he would address the problem of the deteriorated beam ends by replacing them with steel T-ends that picked up the beam loading at the point where the beams showed no deterioration. This dampproofing was one of the greatest challenges Joe faced during the entire restoration.

A project that particularly aroused the curiosity of local preservationists was the excavation of the underground cistern. When the house was built in 1832, the cistern located in the southwest corner of the garden was an important selling point. Before running water came to New York City in 1842, those who did not have cisterns were obliged to haul water from the pumps located on the street corners every two blocks. Joe wondered just how large the Tredwells' cistern was and whether the water had been pumped in the yard or piped into the house. So on the morning of October 11, 1973, with the aid of a stone chisel, hammer, and post-hole digger, he and an assistant uncovered the paving stone containing the manhole cover and began to remove cinder fill and accumulated soil debris. By noon, Joe was able to enter the chamber, which he discovered was a domed space constructed of brick. The next day, they continued removing the cinder fill with a pail and rope. Joe later calculated the cistern would hold approximately 3,800 gallons of rain and ground water and so was no doubt one of the largest cisterns in the city at the time of its installation. He found a pipe leading from the cistern into the northwest corner of the kitchen, answering the question of where the water went.

A hands-on supervisor, Joe Roberto worked alongside the carpenters, electricians, painters, plasterers, and masons throughout the long years of the restoration. He was indefatigable, hauling 90-pound bags of cement; climbing scaffolding to make emergency repairs; constantly consulting with contractors; and problem-solving on every phase of the work. His strength as a restoration architect was his ability to come up with innovative solutions to unusual problems posed by the 140-year-old house, and there were many. Howard Zucker, the painting foreman, described Joe's specs as "thorough, exacting, and meaningful," and he was demanding in their execution. He had a talent for finding craftsmen who could get it right. "I don't think he would have allowed inferior workmanship," said Zucker. [2] John Sanguiliano, who as a student did full-scale drawings of the ornamental basket urns preparatory to their reproduction, called him a perfectionist with an uncanny eye: "He could tell from across the room if something was an eighth of an inch off, and it had to be exact." He did all of this work on a volunteer basis, although the State allowed him to count the value of his professional services as matching funds.

Had Joe not lived and worked in the neighborhood, it would have been impossible for him to do what he did. Typically, he opened the house and consulted with the workmen early in the morning before he began work at New York University and closed the house in the evening after his workday at the university had ended. He was frequently to be found at the house on his lunch hour and almost always on the weekends. He estimated he spent approximately 1,000 hours a year (the equivalent of twenty-five forty-hour weeks) on the restoration of the Old Merchants House.

Joe's commitment to improving his boyhood environment by preserving its historic structures extended well beyond the Old Merchants House. As New York University architect, he encouraged his employer to preserve and restore NYU's landmark buildings and Washington Square Park. He was one of nine architects responsible for the plan to restore the park in 1964. He was also responsible for restoration of "The Row," the imposing group of Greek Revival houses on the north side of Washington Square, which were owned by the university. He oversaw the adaptive reuse of the Paul Manship Studio House, now Deutsches Haus, at University Place and Washington Mews; the rehabilitation of the 21 former carriage

2 Howard Zucker, "Howard Zucker at the Old Merchant's House," *Victorian Homes,* Winter 1986, 66-7.

houses in the mews for faculty housing; the restoration of the façades of the Judson Residence Hall and Tower on Washington Square South, designed by Stanford White; and the installation of a 1790 Damascus Islamic house interior in the lobby and library of the Kevorkian Center for Near Eastern Studies at 50 Washington Square South. The stones and tiles of the ornate interior (purchased in the 1920s by the center's donor) had been numbered, but the key plan for the numbering system had been lost. Using photos taken before the house was demolished, Joe managed to reassemble the gigantic three-dimensional jigsaw puzzle. Other university projects included restoration of the Hall of Fame, the huge Beaux Arts colonnaded memorial to great Americans designed by Stanford White and located on what was then New York University's uptown campus in University Heights in the Bronx, and the adaptive reuse and conversion of turn-of-the-century commercial loft buildings to educational facilities.

As a volunteer architect, he not only restored the Old Merchants House, he drew the plan for the renovation and extension of the Epiphany Library, a branch of the New York Public Library on Twenty-Third Street. His volunteer work also included the restoration of the historic fence around Stuyvesant Park and securing funding for new lighting fixtures on 33 blocks of the Greenwich Village area.

In 1984, Joseph Roberto was named a fellow of the American Institute of Architects in recognition of his accomplishments in historic preservation. Nor did his achievements go unrecognized by the community that benefited from his expertise and dedication. Over the years, he received many awards from appreciative New York City organizations, including the Municipal Art Society, the Victorian Society, and the New York Planning board, to name but a few.

29. Joe Roberto Inspecting the Roof

Joe Roberto hangs onto a dormer window frame of the Old Merchant's House as he examines the condition of the roof. Judging from the way he is dressed, one can assume he was either on his way to or from his job as New York University architect.

12

1972–73 The Divide Widens

From the beginning, a main source of conflict between Joe and The Decorators Club Museum Committee was fundraising. Joe wanted to widen the circle of donors to include persons other than interior designers and architects, groups that had been the main targets of The Decorators Club in the past. In the spring of 1972, shortly after beginning the exterior restoration, Joe consulted a professional fundraiser who agreed that he needed to broaden the base of donors and advised him to send out an appeal. Joe asked Elisabeth Draper, then the president of The Decorators Club, to head a fundraising committee.

Elisabeth Draper was an interior designer of national repute. In 1948, she had been hired by Columbia University to decorate the President's House to make it ready for the new university president, Dwight D. Eisenhower. She not only decorated the Eisenhowers' New York City home but their Gettysburg farmhouse as well. She also decorated the American Embassy in Paris and worked on a number of rooms at the White House and the rooms of Blair House. She was one of the foremost interior designers of what she herself called "that lovely ladies' era of decorating."[1] In Elisabeth Draper, Joe Roberto found a kindred spirit, and he relied heavily on her throughout the restoration.

Under her direction, the Fundraising Committee decided to earmark the funds they were about to raise for the new slate roof, reasoning that people would easily understand this critical need and be likely to respond to it. Joe and Carol put in long hours addressing invitations and writing personal notes to their friends and to a new list of community leaders and professionals. The appeal was

1 Suzanne Slesin, "Elisabeth Draper, Grand Dame of Interior Design, Is Dead at 93." *The New York Times*, July 8, 1993.

very successful, netting more than $9,000. All grant money and matching funds for restoration were deposited in the account of the Historic Landmark Society.

30. ELISABETH DRAPER
Elisabeth Draper, president of The Decorators Club, 1971–73, and a member of the museum board for 18 years, from 1975 to the time of her death in 1993. She worked throughout the restoration of the 1970s, raising funds and helping to reinstall the collection once the structural work was completed.

Cornelia and her Museum Committee knew about this fundraising project, but Cornelia had had surgery that spring and was laid up for a couple of months, so it was understandable they were not involved.

Elisabeth Draper and her committee, encouraged by the success of their first fundraising effort, immediately began planning another—a theater party for the fall of 1972. Because board president Powell Pierpoint wanted the theater party to be an effort of the museum's trustees, they did not think it necessary to tell Cornelia, the Museum Committee, nor even the board of The Decorators Club of their plans. Working through a theater bureau, they arranged for an October benefit to be held at the Uris Theater, featuring the play *Via Galactica,* a science fiction tale starring Raul Julia. In this imaginary universe, everyone was the same color—blue—so there was no racism, and indeed no interpersonal tension of any kind because the citizens of this utopia were equipped with flywheels on top of their heads, which helped them maintain their equilibrium.

Cornelia and the Museum Committee, having been left out of the loop, could have used a few of those flywheels. When they discovered what was happening, they were understandably upset. Why would the Fundraising Committee plan a benefit without informing the board of The Decorators Club and the Museum Committee? Why didn't the names of their board and the Museum Committee appear on the front of the invitation, along with the names of the trustees of the Historic Landmark Society? There were no answers and no apologies. But in the end, everyone bought their tickets, and a good time seems to have been had by all. The benefit made $9,000 for the Old Merchants House restoration, bringing the total raised in just seven months to a little over $18,000. The Fundraising Committee's approach was obviously working.

By 1973, the members of The Decorators Club had become deeply divided about their commitment to the Old Merchants House. Some stood with Cornelia and the original Museum Committee, determined to honor their original commitment; some joined with Joe and Elisabeth Draper and the Fundraising Committee in their efforts to further what they saw as the good of the house; and a large contingent strongly opposed the involvement of The Decorators Club with the Old Merchants House altogether. It was, they argued, too big a project over which they seemed to have insufficient control.

In preparation for the March 1973 meeting of the board of directors of The Decorators Club, Cornelia wrote a memo to herself, declaring her determination to see the restoration through:

1. Purpose of restoration of Old Merchants House from beginning was to give lasting prestige to The Decorators Club.
2. I was appointed permanent chairman of the Museum Restoration Committee and I intend to remain chairman.
3. I intend to see that ten years of work this committee has put into it is recognized and they are given credit.

Addressing the meeting, she announced that it was time to end their bickering. She explained that the Friends account and The Decorators Club restoration account had been separated from the treasury of The Decorators Club in 1967 and funds raised by the Friends and the Museum Committee were specifically designated for the restoration of the Old Merchants House. While The Decorators Club occasionally made gifts to the Museum Committee (they had just donated $2,000 for a deposit towards the reproduction of the parlor carpet), such transfer of funds required approval of the board and were by no means obligatory or automatic. Once again, she passionately made the case, as she had convincingly done over ten years earlier, that the restoration of the Old Merchants House would be a public relations coup for The Decorators Club. "It's time," she argued, to "go forward or forget it!" She made it clear that she was ready to throw in the towel if she did not have the complete support of the board. In the end, they gave her their vote of confidence.

Joe hoped federal grants would continue to provide major funding for the restoration. But there were no guarantees, and of course funds would have to be raised to match every federal dollar received. Realizing that all parties needed to work together to raise enough money to see the restoration to its completion, he wrote to Powell Pierpoint appealing for his help in improving relations with The Decorators Club, but Pierpoint was not inclined to become involved.

The Fundraising Committee launched another appeal in the spring of 1973, which raised over $5,000, and a theater benefit in the fall of 1973 at the Helen Hayes Theater featuring the play *Crown Matrimonial* starring Eileen Herlie and George Grizzard added just over $8,000 to the restoration fund. The fundraisers were on a roll, but once again, for some reason, they had not told Cornelia and her committee of the plans for the theater party. And they did not know, because Cornelia had not told them, that she had planned a fundraising trip to the Newport cottages for that same weekend. Exasperated, Cornelia canceled the reservations for the hotel and the bus.

In the summer of 1973, an incident occurred that served to further damage the relationship between Joe and Cornelia. Upon opening the house on the morning of July 14, Joe discovered that a letter had been slipped under the door. It was notification of foreclosure; the Old Merchants House was now the property of the City. Rental notices, it explained, would be forthcoming. At first, Joe thought it was a joke; somehow John Sanguiliano, the architecture student who was then at work on the drawings of the basket urns, must have somehow obtained an official form and filled it out. But when Sanguiliano vehemently protested his innocence and Joe investigated further, he discovered that the "frontage charges"—representing water and sewer fees—had not been paid since 1966. The Historic

Landmark Society, as a tax-exempt corporation, was not required to pay them, but a filing for exemption had to be made each year, and that had not been done. Repeated notices had been sent to the house and were forwarded along with all other mail to Cornelia, who, not knowing that she needed to file for an exemption, simply ignored them. It took an attorney six months to get the title restored.

1973–75
Hidden Agendas

Towards the end of 1973, with the completion of the exterior restoration in sight, Cornelia and her committee became increasingly concerned about their role once they could start refurbishing the interior. So at the board meeting of the Historic Landmark Society in December 1973, in an effort to clarify their position, Cornelia made the reasonable request that all fundraising projects and plans be cleared with the Museum Committee of The Decorators Club as well as with the board of the Historic Landmark Society. She also requested a joint meeting of those two groups.

But Powell Pierpoint and Joe had a different agenda in mind. They realized that before Cornelia began incurring financial obligations for the restoration of the furnishings, it would be necessary for all restoration funds to be consolidated into one account available to Joe (who, as restoration architect, was solely responsible for the expenses of every aspect of the restoration—including the restoration of the furnishings), but unavailable to Cornelia. This meant that she would have to relinquish control over the funds in the Friends of the Old Merchants House account as well as those in The Decorators Club Museum Committee account.

At the December board meeting, therefore, Powell Pierpoint proposed a committee of himself, Cornelia, Joe, and board member John White, who would serve as chairman, to "set up a new structure for the final phase of the restoration."

White called a meeting immediately. In addition to the four members of the committee, Gerald Cahill, the comptroller of Cooper Union, was present as was Paul McConville, the attorney who was straightening out the matter of the delinquent water charges. The first order of business was a proposal by Pierpoint to change the name of the governing body from the Historic Landmark Society to The Old Merchants House of New York, Inc. The name chosen by George Chapman in 1935 was believed by many to be a misnomer and the source of confusion. The four members of the committee unanimously agreed to the name change.

Pierpoint then asked Gerald Cahill if he would be willing to "maintain all financial records and receive and pay all bills for the Old Merchants House." He agreed. In fact, his role was to be somewhat different from the way Pierpoint presented it. Actually, Joe would continue to pay the bills and maintain the financial records of the restoration and the operation of the house, but he would be "accountable" to Cahill, who would supervise budgeting and accounting records and prepare tax returns.

The office of secretary-treasurer, formerly held by Cornelia, would now be held by Cahill. All existing bank accounts would be closed, and funds formerly under Cornelia's management would be earmarked for the interior restoration and deposited in a new bank account of The Old Merchants House of New York, Inc. Joe, Gerald Cahill, and Powell Pierpoint were to be single signatories on this account. When Cornelia needed money for expenses incurred in planning fundraising events, Cahill would advance her the funds, which would then be deducted from the proceeds when it came time to make an accounting. When she incurred expenses for interior decoration, she was to send the bills to Cahill, who would then actually pass them on to Joe for payment. There is no mention in the board minutes of Cornelia raising an objection to this arrangement. In any case, an objection would have been to little effect since she was greatly outnumbered. In the final order of business, Paul McConville reported that he expected the title to the house to be restored soon. His presence at the meeting may have been simply to remind Cornelia of her failure to pay the frontage charges. At the next board meeting, the provisions agreed upon by this ad hoc committee were approved.

The machinations of the board had been undoubtedly devious. Throughout the restoration, both Joe and the other men on the museum board were reluctant to confront Cornelia. She was a headstrong individual and could be difficult when crossed. It is doubtful she would have given up her fiscal authority without a pitched battle had they used a direct approach. And they did not want to fight with a woman. They had been schooled in the canon of gentlemanly behavior since childhood. The first rule was "you don't hit girls, no matter what."

In April 1974, with the exterior work now all but complete, Cornelia decided to celebrate with a champagne reception at the house. In the wording of the invitation, which billed the reception as an event to honor the Architects Committee, and in her welcoming remarks, she pointedly denied Joe the spotlight, implying that the exterior restoration had been a collaborative effort of all of the architects. The truth was the Architects Committee had acted only in an advisory capacity and was infrequently called upon for advice. If ever a single individual deserved credit for a big job well done, it was Joe Roberto.

Joe may have tried to make this clear when it came his time to speak. Before showing slides demonstrating the work undertaken to date, he reviewed the history of his involvement with the Old Merchants House and pointed out that he was the designated restoration architect "responsible for the program, the cost estimates, the execution of the restoration work and responsible and liable for the approved expenditure of all funds."

But Cornelia was not to be so easily put in her place, as Joe would discover in the days to come.

By May 1, 1974, she still had not turned over the funds she controlled to Gerald Cahill as agreed upon by the board in December. So Cahill arranged to meet her in her Sutton Place apartment on May 29 to discuss financial matters. At that meeting, he extracted from her a promise to turn over the funds to him as of June 30, 1974. For his part, he agreed that the funds she turned over, as well as future funds raised by the Museum Committee of The Decorators Club and the Friends of the Old Merchants House, would be earmarked for the interior restoration. The transfer of funds took place, although Cornelia did not actually close the accounts she controlled, leaving a small balance in each.

In July 1974, the State allotted $35,000 to the Old Merchants House for Phase Three, the interior structural restoration. However, until final approval came from federal officials, no work covered by the Phase Three grant could commence. As it turned out, Joe would have to wait for almost a year while the machinery of the federal bureaucracy slowly ground away. Until then, interior structural work was put on hold.

Meanwhile, assuming that work was about to get underway and their turn was just around the corner, the Museum Committee began making plans. They contacted Berry Tracy, curator of American decorative arts at The Metropolitan Museum, who agreed to act in an advisory capacity. Tracy, who was one of the most respected authorities on nineteenth-century furniture in America, had visited at Joe's invitation early in 1972. Looking ahead, Joe had hoped that Tracy would become involved as a consultant when the time came to make decisions about the restoration of the furnishings. Cornelia was sure that time was approaching. In a report to the Museum Committee, she looked forward to getting down to work, noting "the executive committee of the Museum Committee will set policy and oversee the work."

31. Berry Tracy
Berry Tracy, curator in charge of the department of American decorative arts of the Metropolitan Museum of Art, 1968-81. Tracy served as consultant to the Old Merchants House, advising on the restoration of the interior furnishings.

When Joe got wind of this report, he was alarmed. On July 24, 1974, he wrote to Pierpoint, hoping he would take a stand and help rein in Cornelia:

> As you are well aware, there are many organizational problems affecting the Old Merchants House. These must be resolved so that the restoration can be completed.
>
> Both the Federal and New York State agencies concerned with Historic Preservation define my role as Restoration Architect and hold me solely responsible for restoration, a role that has never been clarified with board members and The Decorators Club....
>
> I must be acknowledged by the Trustees to be in complete charge of

> the entire restoration. This includes sequence and coordination of furnishing work, interior finishes, and garden landscaping....
>
> All orders must be countersigned and committed by me.... I cannot continue the administration of the work performance or its coordination with the funding requirements of the official agencies unless the preceeding operative conditions are established.... Can we discuss this in greater detail?

Even though Joe explicity made clear the urgency he felt and directly asked for a reply, this letter went unanswered for over three weeks. On August 13, he tried again:

> I must have the assurance that the restoration work on the House can continue under my direction. This is a mandatory requirement of the grant. My letter of July 24th set out the condition.

It would be another two weeks before a brief and disappointing answer came from Pierpoint. The letter reads in full:

> I really never considered that there was a question but that you were in complete charge of the restoration. To whatever degree it is necessary to reaffirm that position this letter should serve that purpose. I am sure you know that I recognize the handicaps under which you work. I can only say that the Old Merchants House owes you a debt of gratitude that none of us will ever be able to repay.

But Pierpoint's understanding of the situation was not the problem. Cornelia was the problem, and clearly he did not intend to tangle with Cornelia. Joe was going to have to handle whatever disagreements he had with her himself.

And she was not his only concern. George Chapman had never felt that the board was under an obligation to donate money or raise funds, but Joe felt otherwise. He was dissatisfied with what he perceived as a detached board, unwilling to be actively involved. In May, he had asked Pierpoint to appoint new trustees and had even furnished a suggested list of prospects. By July, nothing had happened to change matters and so in his letter of July 24, he had again urged the appointment of new trustees.

> To function as Restoration Architect has required extraordinary demands on my schedule and energies, all of which I find unsupportable without a working Board of Trustees.... I urge the appointment to the Board of those suggested in my letter of May 29th to you. An active Board, which will give or get money or take dynamic action is imperative.

In January 1975, two of the persons on his list, Elisabeth Draper and Spencer Davidson, were appointed trustees. Federal officials had still not given Joe the go ahead to proceed with Phase Three, which provided for interior structural restoration, but Joe thought surely he would hear from them soon. Still fearing Cornelia would give orders and make commitments without his okay, on March 5, 1975, he again appealed to Pierpoint, pointing out the restoration of the interior had to be carried out in strict accordance with the terms of the grant:

> Excluded are items which have been proposed, such as establishing a kitchen at the rear first floor tea room, the dumbwaiter, ... pergolas, gazebos, wallpaper, etc. I must remain in full responsibility to function. I therefore request designation as Director of the Restoration by the Trustees.

Finally, in an effort to satisfy Joe that something was being done to avert the possibility of Cornelia making unilateral decisions, at the next board meeting, Pierpoint directed that an executive committee composed of Joe, Cornelia, Elisabeth Draper, and Spencer Davidson be set up to administer the interior restoration, all orders to be approved by the entire committee before they were executed. Another committee was not exactly what Joe had in mind.

It wasn't long before it became clear that Cornelia still intended to have it her way. She had written to James Biddle, president of the National Trust for Historic Preservation, hoping to enlist his support for her intention to use the house, once the restoration was completed, as a clubhouse for The Decorators Club. His reply of March 5, 1975, was just what she had hoped for:

> Historic house museums, in general, should be working houses not just static museums. It is important that they be seen and used by as wide a variety of people as possible and not just a museum public. I am pleased to know that is the direction you are heading.

Cornelia sent a copy of this letter to Joe, who, since he had not received a copy of her original letter to Biddle, felt he had been blindsided. If Joe had anything to say about it—and he did—once the restoration was complete, The Decorators Club would not be using the Old Merchants House as a clubhouse.

In April 1975, Joe decided that the bureaucrats in Washington must have lost the paperwork for the $35,000 grant the State had approved back in July 1974. He wrote his congressman, Ed Koch, asking him to look into the matter, and on May 5, he received notice that the Department of the Interior had approved the application and he was free to proceed with the structural restoration of the interior.

He had already applied for the final government grant of $40,000, which when matched, would cover Phase Four: preparing all the surfaces for painting and applying the original colors as determined by a professional paint analysis, as well as restoring and reinstalling the furnishings and laying the carpet. Ever the optimist, Joe thought that if he could just secure the matching funds and receive the go-ahead for Phase Four by November 1975, he could still open the house in the summer of 1976—in time to celebrate the Bicentennial.

14

1975–76
Back to Work

Just as soon as the third grant of $35,000 for Phase Three received final approval, Joe went to work on the restoration of the interior. Workmen converged on the house. They insulated the extension and finished the walls there. They replaced the old electrical system. They stripped windows of accumulated paint and restored them to perfect operating condition with period replacement sash weights where necessary. They repaired broken moldings and replicated rotten windowsills and missing moldings exactly.

In addition, much work remained to be done on the ground floor. In the kitchen, carpenters reinforced the rotted floor joist ends that had been uncovered during the dampproofing work with steel flanges. The numbered and heavily paint-encrusted original floorboards that had been removed were then scraped and sanded and returned to their original locations. The workers then removed, sanded, and reinstalled the wooden wainscoting. Deteriorated brick in the kitchen fireplace was replaced with period-appropriate brick; the chimney were rebuilt; the hearthstones reset. An unsightly duct exhaust along the ground-floor hallway ceiling, a remnant of the small hot air furnace installed by Gertrude Tredwell years earlier, was also removed.

James Biddle found a nineteenth-century soapstone sink, which Joe installed in the northwest corner of the kitchen where the pipe to the cistern was located. He added a period hand pump to show how he assumed the water was delivered from the cistern to the sink.

Even though the elevator the Tredwells installed in the early 1870s had been removed by George Chapman, it had left behind two serious structural problems that Joe now had to overcome. To accommodate the original installation, the stairway from the second to the third floor had been moved forward 42 inches.

As a result, the stability of this stairway was compromised, and over the years it had pulled away from the east wall, leaving a three-inch gap between the stairs and the wall. When Joe uncovered the stair tread on the third-floor turn and discovered the extent of the deterioration, he decided he would have to bring in a professional stair builder to rebuild the stairway. To insure its stability, the stair builder connected the beams on the end of the stairs to the east party wall brick masonry with steel tie rods. He also restored railings and balusters on all floors and installed new balusters turned to match the originals that were missing.

32. Staircase Before the Restoration
The stairway from the second to the third floor was dangerously deteriorated and had to be completely rebuilt.

An even greater challenge existed in the attic above the fourth-floor servants' quarters, where the operating mechanism of the elevator was located. To make room for the cast-iron wheel that hoisted the elevator, a girder had been cut through. The enormous weight of the slate roof bearing down for over a hundred years on the resulting inadequate support had caused walls to crack and beams and floors to sag. To support the load, Joe installed an iron tie rod between the east and west party walls and covered the bolt ends with star plates on the exterior.

33. Elevator Mechanism
The cast-iron wheel that operated the one-person lift installed by the Tredwells for the benefit of the youngest daughter, Sarah Tredwell. Originally located in the attic above the servants' quarters, the wheel is now on display on the fourth floor.

An indication of the pride Joe took in his work is still visible on the fourth floor. When it came time to repair the plaster, he had the workmen leave a framed opening in an interior wall, exposing a portion of the tie rod so that visitors could appreciate his solution to the problem created by the installation of the antique elevator.

In June 1975, another of Cornelia's bus trips, this time to Old Westbury Gardens and the Coe Planting Fields, yielded $1,400, which she turned over to Cahill. The Decorators Club was counting on the agreement made by the museum board that the funds they raised on their own and that were being held by Cahill would be used for the interior furnishings.

34. Fourth-Floor Enframement
In order to show visitors how he had managed to restore support to the roof, Joe Roberto created an opening in an interior wall, exposing the tie rod that runs between the east and west party walls.

Cornelia's husband, Ira Emerich, died in October 1975. Emerich had underwritten the first Kettle Drum party that introduced the Old Merchants House to The Decorators Club in 1962. Throughout the ensuing 13 years since they had first considered taking on the restoration of the Old Merchants House, he had encouraged Cornelia to persevere in her involvement with the house during some discouraging times. She requested that a fund set up in his memory be designated to restore Seabury Tredwell's bedroom. Eventually, contributions to this fund exceeded $4,000.

Cornelia and Ruth Strauss called on Joe to assist them in the first place because they realized he would know what they had to do to save the house from collapse. Still, many of the members of The Decorators Club were not happy with the way things were going. Time and again they had to readjust their expectations. By the beginning of 1976, their dream was still unrealized. However, they reasoned that Joe's work on the interior could not go on much longer, and at last it would be their turn.

And so in February 1976, they shifted into high gear. One thing they no longer needed to discuss was the carpet. Even before structural work on the interior began, the original had been reproduced at a final cost of $9,000 and was stored in the warehouse of Patterson Flynn Martin.

The Museum Committee met twice in Cornelia's apartment where they decided on a course of action. The Costume Committee would inventory and store the costumes. The Furniture Committee would tag the furniture they decided to send to the upholsterers for repair and upholstery, and Cornelia would prepare the purchase orders and schedule the pickup. Belfair Draperies, the firm selected to fabricate the parlor curtains, had been storing the Scalamandre silk since 1967. However, the committee decided not to place an order for Belfair to begin any work until a donor could be found to furnish new fabric for the rear bedroom bed hangings.

On February 16, 1976, Cornelia had 12 pieces of furniture trucked to the firm of P. Nathan with an order for reupholstering, repair, and polishing. She forwarded the order to Cahill and asked him to send it to the restorer along with a check for $914 to cover the down payment. Cahill, of course, passed the request along to Joe, who was shocked that furniture had been removed from the house without his knowledge. Although the trustees' Restoration Committee formed a year earlier by Powell Pierpoint had never met, Joe remembered they had come to an understanding then that all orders would be cleared with the committee. But rather than confronting Cornelia directly, he wrote the check to P. Nathan's and asked Cahill to tell her that the house would still be in too much disarray to receive the furniture by June 1, the promised date of delivery.

When Cornelia received the second-hand message that the house would not be ready by June 1, she called P. Nathan and put the furniture order on hold. Still, she expected she would be able to move ahead with her plans soon and the house would open in the fall of 1976.

But on April 15, 1976, Joe received the demoralizing news that the $40,000 Phase Four grant had been cut in half; the reduced amount of $20,000 would cover only plastering and painting the interior. Somehow money would have to be found for restoring the furnishings and reinstalling the collection. A 1976 Bicentennial opening was now out of the question.

The Museum Committee was aware of the cutback in funding, but they did not as yet fully appreciate the impact it would have on the scheduling. At their May meeting, therefore, they discussed plans for an opening party: a black tie affair with mannequins dressed as the members of a wedding party displayed in the parlors. The bride would wear an 1872 wedding gown, which had recently been given to the museum, and the mannequins representing the wedding guests, wearing the Tredwell dresses, would be grouped around the dining table. Cornelia wrote in her annual report to the club, "At last we are seeing our dreams realized at the Old Merchants House. It has been a long hard struggle . . . but it has been worth it."

In May 1976, a bus trip to three New York City historic houses raised $1,400. By then Cornelia had begun to understand the severity of the financial crisis and suspected that the money The Decorators Club had raised would be used to pay outstanding bills for structural restoration or even for operating expenses. Therefore, to make sure that the $1,400 she collected would be available for interior decoration, she just quietly deposited the proceeds in the Museum Committee bank account, which in spite of the board's directive in 1973, she had not yet closed.

By the fall of 1976, it was apparent to everyone that for the foreseeable future, the work of restoring the furnishings had been suspended.

15

1976–78 Ornamental Treasures

The 1970s marked a low point in the history of New York City. Throughout the decade, the crime rate kept rising, the subways and the parks were unsafe, and the homeless and drug dealers seemed to be everywhere. The population declined as dispirited residents left the city for a better quality of life. It was against this backdrop that the restoration of the Old Merchants House was taking place. With the city on the verge of political and moral collapse, an objective observer could have been forgiven for thinking that it was a pointless effort.

Joe began to fear he might actually lose government grant money for lack of matching funds. He not only needed money; he badly needed a psychological boost. Encouragement was soon to come from an unexpected source.

Richard Jenrette, a cofounder of the Wall Street investment firm Donaldson, Lufkin & Jenrette and a trustee of New York University, had convinced the university to restore the front parlor of 5 Washington Square North, a Greek Revival house owned by the university and used for offices. Jenrette offered his financial support for the project. As university architect, Joe, of course, would be involved in the restoration of the parlor, which was to be used as the office of the Dean of Arts and Sciences and a reception room.

Jenrette was an aficionado of Greek Revival architecture who collected Greek Revival houses.[1] To restore these houses, he enlisted the services of Edward Vason Jones, an eminent architect who was one of the country's foremost authorities on

---

1 Richard Jenrette has collected and restored a dozen houses in the classical American architectural style, which he furnished with antiques of the period. In 2006 he formed the Classical American Homes Preservation Trust (CAHPT). Six of the best of his collection of houses are now owned by the Trust and are open to the public. Jenrette is the author of *Adventures with Old Houses* (2000) and *More Adventures with Old Houses* (2010).

early nineteenth-century interiors and himself a collector of Empire furniture and decorative arts.

35. Edward Vason Jones
At the time Edward Jones was a consultant to the Old Merchant's House, he was directing the first renovation of the US Department of State diplomatic reception rooms. He also worked on rooms at the White House.

Jones was best known for having restored the Adams and Jefferson diplomatic reception rooms at the State Department and for having worked with Clement Conger, White House curator, and two first ladies—Pat Nixon and Betty Ford—to restore the Blue and Green Rooms in the White House. He was also a good friend of Berry Tracy, curator of American decorative arts at The Metropolitan Museum, who had agreed to act as consultant on the furnishings of the Old Merchants House. At Jenrette's recommendation, Jones was engaged to supervise the restoration of the university project.

One day after they began working on 5 Washington Square North, Edward Jones and Joe Roberto made the short trip to East Fourth Street. Jones's first look at the Old Merchants House came at the time when the restoration had reached the precise stage when someone with Jones's expertise could be helpful. The entire structure was now sound; the upheaval inside was over; and the detritus of construction had been removed. Apparently Jones's reaction on seeing the Old Merchants House was everything Joe hoped for. Before long, he had agreed to supervise the restoration of the parlors, and, along with Berry Tracy, to serve as consultant on the reinstallation of the collection. He would eventually bring on a team of the country's most talented craftsmen to return the Old Merchants House double parlor to its original beauty. Like Berry Tracy, Jones contributed his services without charge.

Vibrations from heavy commercial traffic and from the monster printing presses of the Devinne Press Building on the corner,[2] as well as years of water infiltration, had taken their toll on the dramatic ornamental plasterwork of the parlors. Some of the ornaments were loose; others, deteriorated; many were missing entirely. Nevertheless, Jones was impressed by what he saw.

The plaster ceiling medallions of the parlors are unique survivors of the era. A little over five feet in diameter, they are larger than most ceiling medallions found in Greek Revival houses. In addition, rather than being flat as such medallions usually are, their centers of alternating foliate acanthus leaf clusters are recessed into the ceiling. This characteristic creates depth and adds interest, but along with the large size of the medallion, requires an extremely elaborate system of framing and lathing of the central ceiling joists. That the heavy medallions were still in place after 145 years was testimony to the care and skill of the original builder.

The official announcement of Edward Jones as consultant for the Old Merchants House restoration was made on May 3, 1977. Upon Jones's recommendation, Joe engaged David Flaharty, a sculptor and ornamental plasterer who had worked with Jones at the State Department and the White House, to restore the ornamental plaster of the parlors. According to Flaharty, the matching ceiling medallions are "unquestionably among the finest such designs to survive" and in his opinion superior to any composed during the American Classical Revival. This feature alone qualifies the Merchant's House Museum as an extraordinarily important resource for the study of the history of American interior design.

Other members of the team of experts who would restore the parlor ceilings included Odolph Blaylock, master carpenter, who worked with Jones throughout his career in the capacity of right-hand man, and George Peoples, plasterer. Blaylock and Peoples had worked with Jones at the White House. Barney Kubelik, a

2 The 1885–86 DeVinne Press Building still stands at the corner of East Fourth Street and Lafayette Street. It was designated a New York City landmark in 1966.

specialist in decorative wall treatments, would restore the painting of the vestibule walls.

36. **Plaster Ceiling Medallion**
Photography by David Flaharty
The matching parlor ceiling medallions are the ornamental highlight of the Merchant's House Museum. Restored in 1977 by David Flaharty, a member of the team brought on by consultant Edward Vason Jones for the restoration of the double parlor.

In May 1977, David Flaharty began his work. First, he took meticulous measurements and inventoried the missing ornaments. When the time came to replace missing parts, he removed an intact matching piece by inserting a thin chisel between the ornament and the ceiling or cornice and tapping until the bond was broken. Back in his Philadelphia studio, he removed the paint from the ornaments he had brought from the house, then created rubber molds that would be employed to replicate the missing pieces; finally, he returned to the Old Merchants House to affix the replacements in their proper places with plaster. Flaharty's work at the museum, underwritten by Richard Jenrette and the Vinmont Foundation, would continue throughout 1978 and '79. The final result was a perfect restoration of the nineteenth-century artisan's work in all its swirling, rhythmic exuberance.

Preparing the parlor ceilings and the plaster ornamentation for painting turned out to be a slow and costly process. In the nineteenth century, the ceilings were coated with several applications of calsomine, a water-based paint that was easily reapplied when the ceilings became dirty. Subsequently, over the years, coats of oil-based paint had been applied over the calsomine. Because such a mix would eventually lead to paint failure, all of thc paint and calsomine had to be removed by steaming with a wallpaper steamer and scraping by hand to the bare plaster. And that was not all: hidden deterioration made it necessary to remove sections of the ceiling, which were then relathed by Blaylock and replastered by Peoples.

Unfortunately, it was not possible to use the steamer on the cornice and medallion ornaments, for the moisture seeped behind the ornaments and loosened the adhesion. If the ornaments were to be stripped of the layers of paint, there was nothing to do but brush paint remover and/or water over the surface and scrape away the incrustations, inch by inch. In this way the plaster ornaments of the ceiling medallion in the front parlor were cleaned. However, money was so scarce and the process so time-consuming that Joe decided to forego the removal of paint from the ornaments in the rear parlor.

One of the most exciting days of the restoration had occurred four years earlier in October 1973. Ted ten Berge, a painter, had just started removing two coats of blue-grey paint from the vestibule wall. As he worked, he noticed that traces of a decorative treatment of the wall were beginning gradually to appear. It was a Eureka! moment. Joe realized that what they were seeing was a marbleizing effect that had been a popular feature in elegant nineteenth-century homes. Sometime during the latter years of George Chapman's management of the museum, the marble treatment had been painted over because of its deteriorated condition. When Joe discovered an artistic creation lurked behind the paint, he had ten Berge stop his work. Now, four years later, it was time to find out exactly what was there. Excitement mounted among those watching as ten Berge carefully removed two coats of paint and the *faux marbre* treatment gradually appeared. It proved to be a fantasy concept of Siena marble with an imagined graining outlined in blocks 36 inches wide and 16½ inches high. The west wall was in almost perfect condition, but over the years, the east wall had lost half of its original marbleizing.

Kubelik reproduced the base marble effect using oil colors, turpentine, and a solvent. He then used sponges and a variety of brushes to draw in the spidery marble lines, finally applying a protective glaze. So successful was this restoration that it is impossible today to tell the restored sections from the original rendition.

37. Restored *Faux Marbre* Wall Treatment of the Vestibule
The *faux marbre* finish of the walls of the foyer was a popular decorative element in fashionable nineteenth-century homes.

Private foundation grants earmarked for the restoration of the vestibule permitted this work to go forward, but on April 7, 1977, Joe informed Spencer Davidson, who was now president of the board, that unless more funds were forthcoming from somewhere, it would be necessary to cut out all expenses except for gas and electricity or to raise substantial sums from the trustees.

Because of the dire financial situation, Joe turned his full attention to launching a fund drive that he hoped would garner enough response to put the restoration on the road to completion. He asked the trustees to contribute $1,000 or grant non-interest-bearing loans of a greater amount and to do what they could to spread the word of the needs of the restoration to their professional contacts. A press release capitalizing on the involvement of Edward Jones as consultant quoted Jones as saying that if the entire house could be restored to its pristine elegance, "it will rank among the most beautiful in the United States. It will glow like a jewel." The Robertos wrote handwritten notes appealing for support to former donors and friends, and Joe appeared along with Flaharty, Kubelick, and Blaylock in a local NBC TV presentation where before-and-after interior pictures of the restoration were shown. There were no contributions forthcoming from The Decorators Club, and for the second time, Cornelia failed to turn over to Gerald

Cahill the proceeds from a summer bus tour to the Yale Center of British Art and University Art Gallery.

Joe's efforts enabled the museum to limp along for almost a year, but on February 23, 1978, he wrote Spencer Davidson that the restoration was approaching bankruptcy. "By June, there will be no money to pay bills for light, heat, or telephone." Operating costs were running $5,000 a year for utilities, and only $1,000 had been raised in matching funds for the Phase Four government grant.

The winter that year was exceptionally severe. Alternate freezing and thawing of the gutter and the leader caused overflowing onto and through the roof cornice and into the fabric of the building. When spring came, work that had already been done would have to be done again. Joe decided it was time to tell, not ask, the trustees to come up with $1,000 each. If they couldn't meet this demand, he requested they resign so that new trustees could be appointed who could.

16

1978–79
Conflict Resolution

Cornelia continued to ignore the board directive of 1973 to turn over proceeds of her fundraising activities to Gerald Cahill. Furthermore, she and her Museum Committee, believing that the interior would soon be ready for reinstallation of the collection, again began making plans and commitments without Joe's knowledge. They were determined to exercise the authority over the interior restoration that they had assumed would rightfully be theirs once the interior structural restoration was completed.

Elisabeth Draper, who had worked closely with Joe from the beginning on fundraising, and recently appointed board members Carol Roberto and Helen Cranmer[1] joined in an effort they hoped would resolve the conflict between Joe and the Decorators Club Museum Committee. The three women sent Cornelia friendly notes, including copies of articles about the Old Merchants House that had recently been published, and kept her informed of their fundraising projects.

In an effort to forestall the Museum Committee's making unauthorized commitments, Joe sent a lengthy "fact sheet" to all of the members of the committee on May 22, 1978. He outlined his responsibility and accountability to the State just as he had done in his speech to kick off the restoration in 1972. He explained he was responsible for assuring the historical accuracy and quality of craftsmanship of the restoration.

It was logical that interior designers would be involved in deciding paint colors of the interior, and in fact in the spring of 1974, Museum Committee members

---

1 Helen Cranmer was the writer who had interviewed George Chapman 20 years earlier for her book, *Out of This World*. Her first husband, John Erskine, died in 1951; Helen Erskine married William Henry Harrison Cranmer in 1959.

Inez Croom and Mary Dunn had been designated as the Wallpaper and Paint Committee. However, Joe considered the interior wall finishes part of the interior structural restoration and thus his sole responsibility. He was careful to assert in his fact sheet that the colors and method of painting were not a matter of choice but would be determined by a paint analysis that would reveal the original color of the walls. He acknowledged that opening the house would require the participation and help of the members of the Museum Committee. Working together, the first half of the painting phase could begin, but he cautioned, "all decisions, work orders of any kind must be processed through my office." In other words, he was to have the last word on every aspect of the restoration.

Ethyl Alper, a member of the Museum Committee, wrote asking for a copy of the federal grant "so that we can be aware of its provisions." She was probably acting as a surrogate correspondent for Cornelia since Joe and Cornelia do not seem to have been communicating at this time. Ethyl noted that The Decorators Club had been involved with the Old Merchants House for many years and certainly did not intend to do anything contrary to the grant provisions.

Joe complied with her request, but hard feelings persisted throughout the summer of 1978. In July, Cornelia had added the receipts of another bus trip, this time to the New York Botanical Garden and the Hammond Museum and Gardens in North Salem, New York, to her nest egg.

When in October Cornelia abruptly moved to Heritage Village, a Connecticut retirement community an hour and a half from the city, Joe lost no time in trying to get a handle on exactly what commitments she had made to restore the furnishings. His records showed that at her request he had made a payment to P. Nathan for one-third of the cost of reupholstering. Had she made further payments or additional commitments? Had she made arrangements to have the bed canopy frames covered? And what about the curtains for the parlor windows? A trucker's receipt showed that the Scalamandre silk had been sent to Belfair Draperies in 1967. Had they provided estimates for fabrication? Were the curtains perhaps already in production? No one seemed to know, so he dispatched Elisabeth Draper, his wife, Carol, and Helen Cranmer to the firms involved to find out where they stood. They discovered P. Nathan was still holding the 12 pieces of furniture Cornelia had trucked there in 1967 with orders for reupholstery. One-third of the cost had been made as a down payment. Fabrication of the Scalamandre silk now in storage at Belfair Draperies and installation of the curtains would come to $1,500. Furthermore, Belfair had paid $800 to insure the fabric. The Old Merchants House would be billed for this amount, which was now past due. Both of the firms were adamant about having the orders concluded immediately. And still Cornelia retained the bank accounts with the proceeds of her fundraising efforts.

So on December 4, 1978, Joe decided to write a letter to Pierpoint, explaining the situation, appealing to him to prevail upon Cornelia to relinquish the funds still under her control. In an unusual step, he had Carol, Helen Cranmer, and Elisabeth Draper add their signatures to the letter as "endorsers." He asked for a time when they all might meet with Pierpoint personally "to hear firsthand of what we all feel will be a most helpful decision and subsequent action in this frustrating situation." And finally, to emphasize the importance of the request, he had the letter hand-delivered.

Pierpoint's response was to suggest Joe send copies of the letter to the other members of the board; he simply ignored the request for a meeting. So Joe sent off copies of the letter as directed, but the gentlemen trustees were as loath to confront Cornelia as Pierpoint had been.

The resolution of the impasse finally came early in 1979. On January 25, Carol Roberto wrote to Jacqueline Beymer, president of The Decorators Club, informing her that a significant foundation gift would be forthcoming. She enclosed an annual summary of the restoration funds to date and two financial breakdowns. One summarized restoration funds received from all sources since January 1, 1972, and the other was a statement of all contributions, disbursement, and commitments made by The Decorators Club from that time. She noted the last contribution of The Decorators Club had been made in September 1975 and no statement of the profit or loss from the last three benefits had been received. Finally, she welcomed the continued support and participation of all the members of the club. She did not ask that any action be taken, but on February 2, Jacqueline Beymer sent Spencer Davidson a check from The Decorators Club representing the balances of the bank accounts Cornelia controlled. She asked that the money, a total of $7,600, be known as the Cornelia Van Siclen Fund and that it be used solely for the interior furnishings of the house. She concluded by saying The Decorators Club looked forward to resuming its efforts to see the restoration through to the end.

Two weeks later, Joe sent Cornelia a gracious letter citing the "love, time, and continuous effort" she had devoted to the house. He acknowledged how crucial her efforts were in saving the house from complete deterioration and complimented her on her abilities as organizer and director.

After her move to Connecticut, Cornelia continued to be listed as a museum board member. The Museum Committee ceased to be active, although she communicated with them several times by letter. She had hoped to organize a bus trip to see the newly decorated rooms at the State Department. But the next trip was organized by Ruth Lieb, a member of the original Museum Committee.

And in the end, the hanging of the curtains would also be done by others.

17

1979
The Home Stretch

Once it was clear which furniture was in the possession of P. Nathan, and the funds Cornelia had been holding were in hand, Jacqueline Beymer and Elisabeth Draper, working with Carol Roberto, began making plans for concluding the arrangements for the restoration of the front and rear parlors. They met with Berry Tracy, who recommended a few changes to the original orders that Cornelia had given for reupholstering the furniture. Carol Roberto issued the purchase orders on February 13, 1979, signing them "Trustee and Coordinator for the Restoration Architect." As it happened, the money in the Cornelia Van Siclen fund would just cover the amount that would be due for this work, plus the cost of fabricating the curtains, and when the time came, laying the carpet, which was in storage.

The "significant foundation gift" Carol alluded to in her letter to Jackie Beymer was a $20,000 grant from the Vincent Astor Foundation, which over the years had responded to other fundraising appeals. The New York City philanthropist Brooke Astor was an old friend of the museum. Since it was her practice to personally visit the organizations that the Astor Foundation benefited, she had been to the house on several occasions. This large gift provided matching funds for the reduced $20,000 Phase Four government grant. The painting of the interior could now begin.

It had been over 40 years since George Chapman had the parlors painted "colonial yellow," a color that had been his personal preference because he thought it complemented the mahogany furniture. Since that time, professional advances had been made in the field of historic preservation, and the criteria for selecting historically accurate paint colors was based on scientific analysis. In January 1979, Joe turned to color analyst Frank Matero for guidance. Matero performed a chemical analysis and matched what he determined to be the historic colors to available Benjamin Moore paint colors.

38. JACQUELINE BEYMER
Jacqueline Beymer, president of The Decorators Club, 1978-80, and a member of the museum board. Along with Carol Roberto and Elisabeth Draper she was responsible for completing the decoration of the parlors in preparation for the partial opening in November 1979. A public series of scholarly lectures sponsored by The Decorators Club bears her name. The series helps fund the club's scholarship program.

Matero found that for most of the nineteenth century, the walls were one or another shade of white, the same colors, more or less, that they had been at the time of Gertrude Tredwell's death in 1933. And so the walls would be painted white, with one exception. According to Matero's analysis, the original color of the parlor walls in 1832 was a shade of peach that of course had been selected by the builder, Joseph Brewster. Later, in the 1850s, when the Tredwells redecorated and purchased the red-and-gold carpet, they had the parlor walls painted an off-white. But Joe thought that the early peach color was historically significant and interesting, so he had the ground-floor front room painted in the peach color, which he termed "Brewster Pink." [1]

In March 1979 Joe signed a $35,000 contract with Rambusch Decorating that provided for the interior painting of the entire house, and the restoration entered the final stretch. The end was almost, but not quite, in sight.

On April 18, 1979, board president Spencer Davidson notified the trustees of the formation of a new Interior Restoration Committee to replace the inactive Decorators Club Museum Committee. The new committee was designated to direct all phases of the interior restoration, including furniture, window treatments, decorative objects, floor coverings, paintings, and costumes. Edward Jones would serve as the chairman. Cornelia was listed as a member of this committee along with Hope Alswang, Helen Cranmer, Elisabeth Draper, Joe Roberto, and Berry Tracy. They met in May to discuss plans for completing the restoration of the furnishings and the reinstallation of the collection. They hoped the house could be opened to the public in the fall.

Before the painting could begin, much preliminary work was necessary. Three carpenters repaired cracked doors and trim, replaced missing baseboards and floorboards, repaired indoor shutters and shutter pockets, and removed all hardware. Volunteers assisted with the packing and labeling of decorative objects and costumes. Joe had the large furniture moved to the middle of the rooms and covered with tarps. The packed boxes, pictures, mirrors, tables, and chairs were carried to the fourth floor for storage.

The painters began their work on May 2, 1979, but it was immediately clear that it was not to be a routine job. As they began preparing the plaster walls for

---

1 In 2011, advances in scientific paint analysis made an even more accurate determination possible. This analysis revealed that during the 1850s, the walls throughout the house were painted off-white and beige. Neither Matero nor the 2011 analysis found any evidence of wallpaper ever being used in the house.

painting, they soon discovered the house had been hiding a dark secret. When they cut out cracks and blemishes in plaster walls that appeared sound, they encountered hidden deterioration that had not been apparent when Joe rebuilt the exterior west wall in 1972. Before any painting could begin, huge wall and ceiling areas would need to be replastered.

Also, the woodwork throughout the house had been painted in the 1930s with an enamel that did not adhere to the previous coat of paint. The painters were able to lift off this coat easily with their putty knives, but of course, it took time. Afterwards, the woodwork and the columns separating the parlors were thoroughly sanded. Even though the painters experienced unexpected frustrating delays, there was to be no corner-cutting. When it came time to sand the walls, they held a light in one hand as they sanded with the other in order to reveal blemishes not otherwise visible.

When the first walls were finally treated to a coat of primer, the effect was startling. Everyone could see how glorious the house would look when at last a clean, light-reflecting finish replaced decades of grime. It was just the first step, but suddenly it was clear that soon the house would come alive.

Joe was in constant attendance, looking over the workmen's shoulders, making sure his exacting specifications were followed. He directed that only natural hair bristle brushes be used so that the texture of all finishes would correspond to the original nineteenth-century appearance. However, Howard Zucker, the Rambusch foreman, persuaded him to let the painters use rollers on large areas and then to level off the paint with a bristle brush. The rear parlor was the first room to be finished, on August 3, 1979. Joe's log entry for that day noted his satisfaction and demonstrates how important it was to him that every detail of the restoration adhere to historical accuracy.

> The rear parlor is finished. The ceiling is flat, with swirl brush marks to simulate calsomine brush patterns.The walls are finished with vertical brush strokes. The texture of the brush strokes gives a two-tone effect, very different from the flat alkyd paints that we have grown accustomed to seeing.

The painters finally finished their work on October 10, 1979, five months after they had begun. The Rambusch truck came and carted away eleven ladders, nine extendable platforms needed for the scaffolding, and what seemed to be acres of drop cloths.

Seven severe winters had been hard on the house since the restoration began. While the interior painting went on, other workmen were busy repeating work that had been done earlier. New cracks in the masonry of the west wall were causing water seepage at the fourth floor level. Joe had the outside shutters repainted and rebuilt one that was blown off in a storm. He also repainted the cornice and 12 exterior front windows, repaired the front door enframement, and rehung the door. The treads of the cast-iron fire escape had been repaired in 1972, but now many of them were badly corroded and needed to be replaced. New window guards were installed, and painting of the ornamental iron basket urns and fencing was an annual occurrence.

On several occasions, an emergency situation required immediate attention. When Joe discovered that water was leaking into the sidewalls from roofs of the garage buildings on both sides, he quickly made arrangements with the owners of those buildings to repair the garage roofs along the sidewalls, with the museum footing a portion of the cost. One Sunday in May, the third-floor toilet tank in the vacant caretakers' apartment leaked, flooding the second-floor front bedroom, necessitating a new toilet and flooring for the caretakers' apartment bath.

But the most discouraging setback occurred in September 1979 on Labor Day. Workmen were clearing the rooms on the ground floor, preparing for the painters, when they discovered the carpet in the front room was mildewed. When they picked up the carpet, they found the underlayment mildewed as well. It followed that the linoleum under the underlayment was disintegrated and the floorboards rotted. Joe realized the complex drainage system he put in place in the kitchen in 1974 would fix the problem, and he immediately installed a similar system in the front room. There was no need this time to wait to see if it would work.

The museum board agreed arrangements should be made for a resident caretaker couple. The ideal candidates for the job seemed to be Hope Alswang and Henry Joyce. Hope was a curatorial assistant of decorative arts at the Brooklyn Museum and Henry was director of exhibits at the Bronx Museum. Their museum credentials would not only fit them for the demands of curators of the Old Merchants House but, it was hoped, impress potential donors. The couple would live in the caretakers' apartment on the third floor, which by this time was in need of a complete refurbishing, including a new kitchen and an electrical upgrade. By September 1979, their apartment was ready and the new caretakers moved in.

It is not likely Joe would have forgotten that the three-year period of the $20,000 Phase Four grant approved in 1976 ended in April 1979. However, when on June 8, a month after the interior painting began, he received a request from the authorities demanding documentation of expenditures incurred by the April 12 deadline so that they could make final payment, he was shocked. Apparently, he thought an extension would be granted when he submitted his final bills with an explanation of how the hidden deterioration of the plaster had slowed his work. But he was wrong. He immediately dispatched a telegram to the commissioner in charge of administering the grants, explaining the circumstances. The commissioner bumped the request up to the director of the service, who kicked it back down through the chain of command. No requests for extension received after the project completion date would be allowed. Roberto asked his congressman, William Green, to intercede, but the bureaucracy could not be budged. Joe had already received partial payment for work done in Phase Four; the denial of the final payment meant that he would forfeit over $6,000 of the originally allotted $20,000—a disappointing conclusion to the seven-year relationship of the Old Merchants House and government bureaucrats.

Joe would now have to come up with more funds to pay for the painting as well as completing the reinstallation of the collection. Congressman Green suggested that the National Endowment for the Arts would be a logical agency to approach for funding, so Joe immediately composed a proposal and a request for $21,000. Betty Whitman, a long-time volunteer, was hired to head a fundraising effort that included the designing and printing of 7,500 promotional brochures. Once again, the Robertos sent handwritten appeals to friends and former donors. The new curators did their part, securing a $3,500 donation to pay for the renovation and maintenance of their new apartment and making a $2,000 personal contribution.

With a fall opening of the entire house now out of the question for lack of funding, the board decided to have a partial opening of the double parlor only. The furniture had been restored; the carpet had been ready for years and was in storage, as was the silk reproduction fabric for the parlor curtains. Elisabeth Draper, Jackie Beymer, and Carol Roberto would handle the details. They made a decision to have the parlor curtains fabricated by Thomas de Anglelis rather than

Belfair, the firm that had stored the fabric for so long. They set November 29 as the date for the opening.

The Decorators Club Museum Committee had long ago made the decision to open the pocket doors and treat the front and back parlors as one dramatic space. They planned to permanently exhibit the rear parlor with the dining table in the center and the Duncan Phyfe chairs arranged around it. Even though the silk fabric had been used only for the front-parlor windows in the nineteenth century, the Museum Committee had asked Scalamandre to reproduce 100 yards of the fabric, more than enough for the windows in both the front and rear parlors and the reupholstery of the sofa.

Apparently neither Edward Jones nor Berry Tracy questioned this interpretation of the double parlor. However, the impression it conveyed was somewhat misleading. Subsequent research has shown that during the nineteenth century, in homes like the Old Merchants House, the rear parlor was primarily a family sitting room, and the pocket doors separating the two rooms were almost always kept closed. Only occasionally, when the family entertained at a formal dinner or reception, would the doors be opened and the space treated as one large room. For a formal dinner, the dining table would be extended and dining chairs placed around it. For a reception, the table would be removed and chairs placed around the edges of the room, clearing the way for the circulation of guests.

On November 1, workmen arrived to install the reproduction carpet. But it was not simply a matter of rolling out a carpet that had been made to measure. Before 1877, when the first broadloom carpet was introduced, carpeting was woven in strips that were sewn together as they were laid on the floor. Like the original, the strips of the Old Merchants House reproduction carpet are 27 inches wide. However, after installing the border, which had been woven separately, rather than sewing the strips together on the floor, which had been the nineteenth-century practice, the workmen tacked them in place.

On November 12, the silk curtains were delivered, and at long last, they were hung at the parlor windows, an event the members of The Decorators Club had been anticipating for 12 long years.

Under the direction of the new curators, Hope Alswang and Henry Joyce, volunteers brought the lamps, china and glass, fireplace accessories, mirrors, and paintings down from the fourth floor where they had been stored. They cleaned and polished everything before putting it in place for exhibit. The gas chandeliers were cleaned, and a professional furniture restorer supervised the polishing of the furniture.

39. The Front Parlor, 1980
Photography by Jaime Ardilles-Arce
The front parlor as interpreted by The Decorators Club in 1979, when the parlor floor was first opened to the public.

On the day before the opening, Spencer Davidson brought the wine, Carol Roberto brought the cheese and crackers, and Henry Joyce created arrangements of flowers and ferns. At last, they were ready. The total cost of the restoration of the double parlor was $39,000.

The next day 54 donors and trustees attended the partial opening of the Old Merchant's House and marveled at the transformation. Cornelia Van Siclen and Elisabeth Draper, as members of the Old Merchants House board, were present, but no other members of the original Museum Committee were there to receive congratulations. They had not been invited. Their turn to view the result of their work would come later at a private showing.

18

1980–81
Finishing the Job

In May 1980, the board decided to eliminate the position of live-in caretaker, and Henry Joyce and Hope Alswang were given their notice. The third floor would be given over to office space.

The objective for 1980 was the restoration of the bedrooms and the ground floors and the final reinstallation of the collection. In February 1980, Joe learned the Old Merchants House would receive the $21,000 grant from the National Endowment for the Arts he had applied for in June 1979, and the following month, the Vincent Astor Foundation awarded the house another grant in the amount of $44,000. The Interior Restoration Committee that replaced the Decorators Club Museum Committee had met for the first time in May 1979. Assured of funding, they could now move confidently ahead.

They met on June 12, 1980, to finalize plans. Members of the committee as it was originally constituted in 1979—Edward Jones, Berry Tracy, Elisabeth Draper and Joe Roberto—were present, as was Carol Roberto, who was not a member of the original committee. Helen Cranmer and Hope Alswang were no longer members of the committee; Cornelia Van Siclen sent her regrets. At this long meeting, final detailed decisions concerning every aspect of the restoration of the second-floor bedrooms and the stair hall were made.

A remnant of a nineteenth-century floral-patterned chintz, which had been discovered in the attic, provided a plan for the restoration of the rear bedroom. Brunschwig & Fils offered to donate an authentic reproduction of this nineteenth-century fabric, which would be used for window curtains and bed hangings on the large four-poster. White cotton hangings and a shirred top would be installed on the small bed.

For the floor covering in the rear bedroom, Tracy recommended a China straw matting in keeping with the pretty, summery look of the floral chintz. The decorative style of the bedrooms would reflect appropriate gender distinctions, underscoring the fact that in a nineteenth-century home like the Tredwells, it was customary for husband and wife to have separate bedrooms. In addition, the selection of the chintz, the straw matting, and the white cotton would suggest the way the room might have looked in the summer if the decorations were seasonally adjusted, as Berry Tracy informed them was sometimes the case in the nineteenth century.[1]

The bed hangings George Chapman had installed in the front bedroom in 1935 were usable. New fringe, tassels, cords, and canopy would have to be provided, but after a careful dry cleaning, these bed hangings would be reinstalled. Since the original wool swags that had been on the bed in the rear bedroom were still usable and matched the hangings in the front bedroom, they would be fashioned into valences for the front-bedroom windows. At Berry Tracy's suggestion, a Scalamandre reproduction of an 1830 ingrain carpet was adopted for use in the front bedroom and the adjoining hall room.

In addition, at this meeting, the committee agreed to send some furniture out for repair, and a red-wool carpet, which matched the border of the parlor carpet medallions, was chosen for the halls and stairways.

Joe took it upon himself to design the kitchen. He placed a cast-iron stove manufactured by Abendroth into the fireplace to accurately represent the method of cooking at midcentury when the open hearth had been abandoned in many homes, including the Tredwells, in favor of the more modern stove.[2] Along the east wall, a large cabinet Joe had made from old doors and other old wood displayed the Tredwells' English ironstone tureens and platters. The soapstone sink and a hand pump were already in place, having been installed when Joe excavated the cistern in 1973. The Tredwells' kitchen worktable and a donated pie safe and butcher block completed the picture. Visitors would enter through the ground floor. The front room, which was to serve as a reception area, would not be interpreted historically.

Joe had the original paths of Manhattan schist in the garden area reset and

1 Since the Tredwells spent their summers for the most part at their country property in Rumson, New Jersey, they would probably not have changed the decorative style of the bedrooms seasonally.

2 A ledger kept by Seabury Tredwell lists the purchase of a range in 1856 for a cost of $37.76.

leveled. Peter Malins, Chief Rosarian of the famous Cranford Rose Garden at the Brooklyn Botanic Garden, supervised the plantings along the perimeter of the garden and the planting of pachysandra in what had probably been two grass plots between the paths in the nineteenth century. The higher level at the rear of the garden, which provides seating ledges, had been created by St. John Simpson, a landscape architect, in the 1960s.

The new chintz bed hangings and curtains for Eliza Tredwell's bedroom would not be ready until the spring of 1982, but by October 1981, the restoration was almost complete, and three floors were finally opened for viewing. There was no official final public opening, but the Christmas party that year was a particularly festive occasion featuring the famous actress of silent movies, Lillian Gish, reading "The Night Before Christmas."

During 1980 and 1981, the Old Merchants House received a lot of media attention. However, the importance of the role The Decorators Club had played was seldom acknowledged. An illustrated, seven-page article in *Americana Magazine,* for instance, noted only that "the [Old Merchants House] board has received more than a dozen sizable donations, as well as free assistance from professional organizations such as The Decorators Club." [3]

In 1973, the landmarks law had been amended to provide for the landmarking of interiors. In December 1981, the Old Merchants House was awarded this distinction. At the time of the interior designation, there were 839 individual landmarks, but only 40 had achieved interior landmark status. The designation covered the ground floor, the parlor floor, except for the rear extension, and the second floor, including fixtures and interior components of those spaces. The designation report described the Old Merchants House as a "unique document of its period that shows with unrivaled authenticity how a prosperous New York city merchant and his family lived in the mid-nineteenth century."

Jacqueline Beymer, as the past president of The Decorators Club and a member of the museum board of directors, spoke at the hearing. She reviewed the role The Decorators Club had played over the 18 years since they had taken it on as their major cause and affirmed their continued interest in preserving and present-

3 Roger M. Williams, "Old Merchants House," ***Americana,*** 9 (September/October 1981): 66–72.

ing "a historical house that has in effect never been changed, only restored with museum-caliber accuracy."

Now to make structural changes to the Old Merchants House, inside or out, would quite simply be against the law.

40. The Front Bedroom, 1981

Photography by Joel Greenberg

The bed hangings and the Tredwells' gessoed finials installed by George Chapman were retained by The Decorators Club in the front bedroom, and the swags of the bed valance that had been on the bed in the rear bedroom were used on the front-bedroom windows (see figures 12 on page 17 and 13 on page 18). The carpet is a reproduction of an 1830 design.

41. The Rear Bedroom, 1981

Photography by Bruce Buck

For the rear bedroom, a reproduction of a nineteenth-century floral-patterned chintz was fashioned into a bed valance of swags and jabots, bed hangings, a dust ruffle, and window curtains. The reproduction fabric was named "Tredwell Garlands." The ornate gilded finials, which were owned by the Tredwells, were found in the attic by George Chapman.

42. The Kitchen, 1981

Photography by Madeleine Doering

A modern stove had been removed sometime in the twentieth century. The double-oven, coal-burning Abendroth stove installed by Joe Roberto dates to the 1840s and is typical of the cookstoves of the period.

43. Brooke Astor and Lillian Gish

Photography by Karen Radkai

Lillian Gish, silent movie actress (left) and Brooke Astor, philanthropist, in the front bedroom after the Christmas party in 1981 when the actress entertained guests with a reading of "The Night Before Christmas."

19

1981–86
From Restoration to Presentation

From the time the parlor floor was opened to the public in 1979, Joe and Carol had assumed the responsibility of welcoming visitors to the museum on Sunday afternoons from 1:00 to 4:00 p.m., and volunteers from The Decorators Club offered guided tours by appointment during the week.

Now that the work of the restoration was over and all three floors open for viewing, the Robertos tried to find their footing. If the museum were to be taken seriously as a historic house museum capable of attracting the funds necessary for the maintenance of the then 147-year-old building, they would have to expand the hours open to the public, offer educational programs for both elementary and high school students, and provide a venue for lectures and events of interest to the general public.

Beginning in the winter of 1981, a Development Committee chaired by board member Sam Burneson concentrated on raising funds for a director's salary, operating expenses, programming, exhibitions, and once again waterproofing the west wall. At the end of 1982, Burneson was able to report that considerable progress had been made, but there was a long way to go. Joe assessed the situation.

They had enough money to hire Patricia Hildebrand as a full-time administrative assistant to help them manage the museum. They also hired Aline Hill-Ries, a former teacher with experience in museum work as a part-time educator who would write grant applications for school and public programs. James Murphy, a fundraising consultant, was engaged to analyze the financial situation and write a report that would make recommendations for action.

But by the fall of 1983, it was clear the museum was not getting the traction Joe had hoped for, no matter how hard he tried. And so, like Chapman before him, he began an effort to turn the management of the museum over to others.

44. Carol and Joe Roberto
The New York University Archives
Joseph and Carol Roberto shortly after the restoration was complete and the entire museum was finally opened to the public in 1980. Carol Roberto, an interior designer, worked with Elisabeth Draper and Jacqueline Beymer, who were members of The Decorators Club as well as museum board members; and Helen Cranmer, a museum board member, reinstalling the collection.

Three years earlier, shortly after the opening of the parlors, Joe had written to James Biddle, president of the National Trust for Historic Preservation, suggesting the board would be willing to deed ownership of the house to the trust and work toward establishing an endowment for upkeep. Biddle, who was a friend of Joe, was encouraging, but shortly thereafter he left the trust, and Joe let the matter drop.

Now he reconsidered the idea. He and Carol were both in their seventies. Perhaps in the interest of the long-term survival of the museum, it was time to relinquish control and hand over responsibilities to an organization that could be trusted to provide thoroughly professional and sensitive stewardship of the house he had so lovingly and skillfully restored. Once again he approached the National Trust.

In the spring of 1984, the Old Merchants House board members met with trust officials who outlined major changes they would require if an agreement were to be reached: First of all, the board would have to hire a staff including an executive officer, an administrative assistant, and a maintenance person. The role that the Robertos would play in the management of the museum was explicitly defined: "Both should become part of the governing board of directors . . . whose individual votes do not dictate the day-to-day operations of the institution." The trust officials who had analyzed the museum operations clearly believed the museum was too closely held by the Robertos and hoped their energies would be channeled "into a team player spirit" so the executive officer could "carry out the program with the least encumbrance." The board should be more active, particularly in the role of fundraising and making policy. Those who were not ready to work as fundraising activists should resign. A master plan should be put in place to effect the financial stability of the museum and the development of its programs.

But the most demanding requirement was that an endowment of $2 million be raised within two years. The trust concluded this sum would be necessary to provide the funds to operate the institution properly and professionally. Considering the difficulty in fundraising the museum had faced in the past, a $2 million goal must have seemed daunting

Nevertheless, at a special meeting of the museum board on June 14, 1984, five members were appointed to a committee to arrange the details of an agreement deeding the Old Merchants House to the National Trust. Besides the Robertos, board members Spencer Davidson, Peter Krulewitch, and Merrikay Hall, an attorney, agreed to serve on this committee.

For the next two years, trust officials visited the museum periodically to train the staff in professional museum management. Both of the Robertos and Aline Hill-Ries churned out grant proposals for a management team, educational programs, and restoration needs. They scheduled walking tours, themed tours of the house, lectures, concerts, exhibitions, receptions, and parties. And they began renting the house for seated private luncheons for 12 in the front room of the ground floor, or buffet dinners for larger groups of up to 40 people utilizing both the ground-floor front room and the historic kitchen. Elisabeth Draper took on the task of personally placating those members of The Decorators Club who had not understood the role Joe was required to play as restoration architect, accountable to the grant authorities for all aspects of the restoration. Joe sent a letter to The Decorators Club thanking them for their past involvement and encouraging them to take an active role in fundraising.

The most interesting effort to raise funds occurred in September 1984 when a crew arrived to film a short documentary, *Teedie from the Childhood of Theodore Roosevelt,* focusing on the boyhood life of Theodore Roosevelt.[1] Even though Roosevelt's boyhood home was located nearby on East Twentieth Street, according to the publicity release, the producers preferred the plain white walls of the Old Merchants House for filming the interior scenes.

Since the museum was then open to the public only on Sunday afternoons, the film crew had the run of the house during the week. All of the rooms, and the garden as well, were pressed into service, one of the servants' rooms being used for the childhood "Roosevelt Museum of Natural History." The museum's collection of lamps and paintings and furniture was rearranged to suit the needs of the storytelling. To augment these objects, the crew brought in an unusual assortment of props: a stuffed buffalo head, taxidermy equipment, a nineteenth-century teeter totter, parallel bars, clocks, candles, pictures, inkwells, and even a live mouse. Sickly Teedie slept in the Tredwells' sleigh bed and seemed to throw up in the Tredwells' commode. The project yielded a profit of $2,600.

In spite of their best efforts, in October 1985, the board finally faced the fact that an endowment of $2 million was an unattainable goal. The two-year effort to get the Old Merchants House accepted as a National Trust property finally had come to naught.

The Decorators Club continued to support the museum throughout the 1980s. At their Christmas party in 1980, they presented the museum a check for $6,000, the proceeds of a benefit concert featuring Robert Herring, baritone. During 1984 and 1985 alone, the Club raised $10,000 for the endowment fund and passed on a bequest of $13,625 left to the museum by one of their members. Past presidents of the club Jacqueline Beymer and Elisabeth Draper served on the museum board during the 80s as did Sarah Tomerlin Lee, who would become president of the club in 1987. An acclaimed figure in the world of design, she was the head of Tom Lee, Inc., a New York firm noted for its hospitality design and its expertise in re-

1 The film, "Teedie from the Childhood of Theodore Roosevelt, Parts 1 and 2," can be viewed on YouTube.

storing historic properties. She supervised the design or redesign of over 40 major hotels, including the Helmsley Palace, the Parker-Merridien, and the Hilton Hotel in New York City. She was also active in the preservation community, having co-founded the New York Landmarks Conservancy[2] and served on its board for over 20 years.

In spite of the efforts of Elisabeth Draper and Joe himself at reconciliation, some members of The Decorators Club still resented the way the restoration had come to pass. Certainly, it had not worked out the way Cornelia Van Siclen and her Museum Committee had envisioned in 1962, even before the beginning of their official adoption of the house in 1963.

For nine years they were indefatigable in raising funds to make emergency repairs and pay the utility bills as they made plans for restoring the interior. They garnered many significant in-kind donations for the museum, and when the house finally threatened to fall down around them, Cornelia Van Siclen and Ruth Strauss had the good sense to realize their limitations and call on Joe Roberto for help. When division among members of the club threatened to jettison the project in 1973, Cornelia had exercised her full powers of persuasion to keep them on board. During the long hard years of the structural restoration they had stood by, frustrated and disheartened, for it was all too often clear to them that their plans were considered less important than the vital structural work necessary to keep the building from collapse.

Yet in the end, it is the work of The Decorators Club that in large part gives the museum its unique authentic appeal.When they first took over the management of the museum in 1963, they could easily have discarded the Tredwells' tattered curtains and threadbare carpet, replacing them with period appropriate versions. However, they wisely chose to respect the Tredwells' taste, reproducing the parlor window curtains and carpet. They reshaped and patched the original bed hangings for the front bedroom and chose a remnant of fabric that had belonged to the Tredwells for the reproduction of the bed hangings in the rear bedroom. During the nine years when they were solely in charge of the restoration, they were careful to keep all of the contents of the house intact. As a result of their efforts and early reserve, today visitors are able to connect to the past through the home of

2 The New York Landmarks Conservancy is an organization committed to the preservation of historic buildings and neighborhoods in New York City. Early successes included saving the U.S. Custom House and the Fraunces Tavern block in lower Manhattan, and restoring the Church of St. Ann and the Holy Trinity in Brooklyn. The Conservancy administers financial and technical assistance to owners of historic properties.

a real family—in a way that they could not if the museum were simply a generic example of a home of the nineteenth century.

45. Sarah Tomerlin Lee
Beyer Blinder Belle
President of The Decorators Club, 1995-97 and member of the museum board, 1986-90.

The clubhouse they had dreamed of had somehow slipped from their grasp, but they had made possible an outcome of even more significance, the preservation of a unique historic site that demonstrates the life of the wealthy merchant class from 1835 until the end of the Civil War. The museum is one of the city's most important historic documents. If there had not been The Decorators Club, today there would be no museum.

On February 12, 1986, the club held a major fundraising event at the museum—a party to honor Cornelia Van Siclen and Elisabeth Draper who, each in her own way, had been instrumental in saving the house.

They both received The Decorators Club's medal of honor, and Cornelia in her first appearance at the museum in seven years, made a short speech. She remembered the day in 1962 when she saw the house for the first time and was inspired to convince The Decorators Club to take on the Old Merchants House as their project. And she reminded them of the day in 1965 when schoolchildren marched through the streets in support of the landmarking of the Old Merchants House and she had welcomed them on the front steps.

Some members of The Decorators Club felt they never received the credit they deserved. They were right.

20

1986–88
Prelude To Loss

In 1985, the museum had received a grant of $5,500 for waterproofing the east wall and repair of the cornice from Community Board #2. That work was completed in December. But water seepage from the west wall into the ground-floor family room had continued to worsen, eroding the mortar between the bricks of the party wall and causing deterioration of the interior plaster and paint finishes. As the new year approached, the need to restore the west wall had become urgent. Only $4,200 had been raised for this purpose, so in January1986 Joe again turned to Community Board #2, made an application for a $30,000 grant, and crossed his fingers.

After a very long wait of 15 months, the grant was finally approved in April 1987. Because the nature of the water seepage was in Joe Roberto's words, "obscure," he called on a *pro bono* committee of experts—architects, preservationists, and a chemistry professor—to analyze the problem and make recommendations. That the problem was serious there could be no doubt. The chemistry professor noted that he had not seen as much water damage inside a building since the great Venice flood of 1966. The committee recommended that the waterproofing of the foundation wall be undertaken as soon as possible while further study of the upper portion of the west wall was under way. What the committee proposed involved digging a trench sixty feet long and five feet deep alongside the entire west wall. Furthermore, they recommended that rather than working from the interior of the Old Merchants House, the waterproofing be done from inside the building next door, which was owned by the Hittner Trucking Company and used as a garage. This way the waterproofing would be on the "outside" of the wall, thus protecting the masonry.

It seemed to Joe highly unlikely Hittner would allow a 60-foot trench to be dug in what was a repair shop for his trucking operation. But he asked—and to his surprise Hittner agreed. If the work crew could do the waterproofing without interruption of the truck repair, they were good to go.

On June 16, 1987, the low bidder, JLG Construction, began work. They dug the trench along the entire foundation wall, applied two coats of masonry cement and sand finish, and finally two coats of "Thoroseal," the same highly moisture-impervious cement-based waterproof coating that had been applied to the upper portion of the west wall in the spring of 1972, shortly after the restoration had begun. When the foundation wall was dry, they applied cold asphalt flashing cement over the Thoroseal and followed up with an asphalted muslin membrane. Next they built a concrete block wall on the outer face of the foundation wall for supplemental support and waterproofing, and finally they filled the trench. With this major repair, Joe hoped that water infiltration from the west was now solved.[1]

Repair of the interior damage to the plaster caused by infiltration of water through the west wall into the ground-floor family room would have to wait however, for the Old Merchants House was about to face another challenge—one that would threaten its very existence.

The three nineteenth-century rowhouses to the east of the Old Merchants House had long ago been reduced to two stories and combined into one large building that was also owned by the Hittner Trucking Company and used as a garage. The garage building was sandwiched between the Old Merchants House and another landmarked site, the Skidmore House, a Greek Revival home built in 1845.

In July 1987, just as the west wall repair was coming to an end, Hittner finalized the sale of this garage building on the east to the Goldman/Minskoff organization. Trouble began immediately. The new owner left the unoccupied premises open, unguarded, and illuminated 24 hours a day. The former garage became a round-the-clock neighborhood supermarket for the selling and buying of crack cocaine and other drugs.

On September 26, thieves broke into the museum through the hatch of the roof scuttle, made their way down the enclosed stairway from the servants' quarters,

1 Unfortunately, the 1987 repair did not solve the problem. The west wall has continued to require periodic major repair and reconstruction.

forced open the bolted door at the foot of the stairs, and descended the two long stairways to the parlor floor where they disabled the alarm. The alarm sensors, of course, should have been triggered in the amount of time it took the thieves to get to the alarm box, leading Joe to register a complaint with the protection service. All of the office equipment was stolen, including a calculator, the copier, two typewriters, an answering machine—even the telephones. In addition, the thieves made off with $1,000 worth of tools, an antique mirror, a silver candlestick, and three clocks. Roberto suspected that the denizens of the crack house were involved, but the case was never solved.

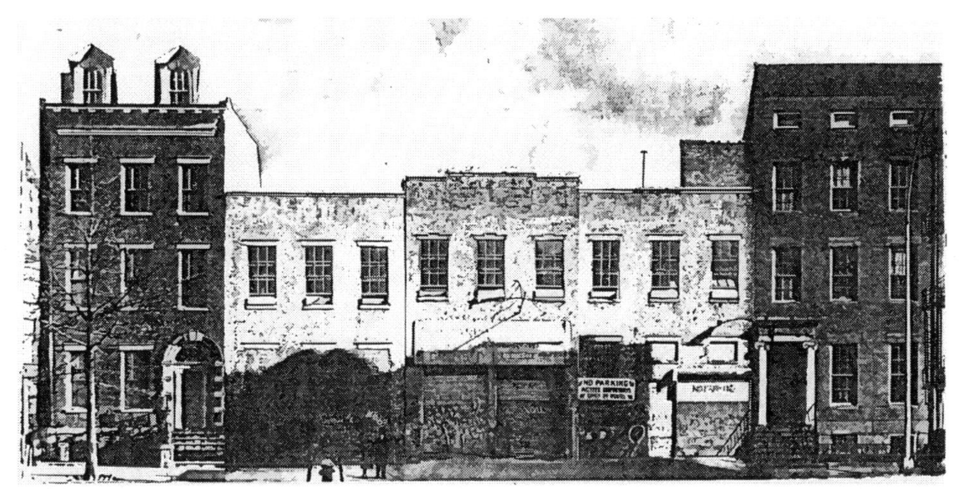

46. Fourth Street in 1988

Between the landmarked Old Merchants House and the landmarked Skidmore House, the three 19th-century buildings that had been reduced to two stories and combined to form a garage building were slated for demolition in 1988. Today the City owns the lot on which the garage stood. It was acquired by eminent domain and affords access to the third water tunnel.

Appeals to the owner, to the mayor, to Community Board #2, to the NoHo Neighborhood Association, and to the Landmarks Preservation Commission finally succeeded in getting the building next door sealed on November 1. The squatters were evicted, but now the new owners announced their intention to demolish the building and use the space as a parking lot until they could get the area rezoned for residential use. They intended to eventually erect a six-story apartment building on the site.

Joe was concerned because the proposed demolition of this building would expose the porous east party wall of the Old Merchants House to the elements and to subsurface water seepage. He estimated it would cost the developer $40,000 to $50,000 to undertake the necessary precautionary steps to protect the museum's east wall during the proposed demolition.

On November 18, 1987, Joe sent a certified letter to Richard Rosen, vice president of the Minskoff[2] organization urging that no demolition take place without assurance that the structural stability of the party wall be unimpaired and that the wall be adequately weatherproofed. He also requested that he and Community Board #2 be notified by certified mail when Minskoff applied for a demolition permit.

Two days later, in a letter dated November 20, 1987, Rosen assured Joe that demolition would be performed in a "thoroughly professional manner with regard to the adjacent landmark properties."

But by December 8, they were erecting a sidewalk bridge, signaling that demolition was eminent, even though Joe had not received any notification that Minskoff had applied for a demolition permit. He realized then that engineering and legal consultation would be necessary to make sure the demolition would take place in a responsible manner. In an emergency memo, he appealed to board members to pledge $1,000 each to meet the need for the estimated $7,500 for consulting fees. On December 11, Councilwoman Miriam Friedlander wrote the Landmarks Commission, asking that they expedite a meeting between Joe and Minskoff representatives to discuss details of the proposed demolition procedures.

The engineering consultant Joe hired began work immediately. He inspected the site and outlined procedures that should be taken in demolishing the building. Joe added further specific recommendations of his own, including the building of an underground retaining wall in addition to waterproofing procedures. On December 23, 1987, he, Carol Roberto, and attorneys and engineers representing the Minskoff organization and the museum, as well as representatives from the Department of Buildings, met to discuss Joe's recommendations. Again the Minskoff representatives declared they were committed to being a "good neighbor," but they explained that they did not have the authority to accept the recommendations. Those would have to be approved by the executives of the Minskoff organization. Their engineer promised to pass them on.

---

2 Sol Goldman, an original partner of the Goldman/Minskoff organization, died on October 19, 1987.

On Christmas day, in an interview with David W. Dunlap, metro reporter for *The New York Times,* Robert Kandel, the Minskoff attorney, was quoted as saying, "We will be taking these buildings down the way they went up, brick by brick, because of our desire to be a good neighbor." He went on to state that the common brick façade would be divided cleanly by masonry saw rather than chopped apart.

When a week later Joe still had not received written approval of the proposed recommendations for demolition from the Minskoff organization, he had the museum's attorney send a hand-delivered letter to the Minskoff attorney insisting that a written agreement to abide by the recommended procedures be sent prior to demolition. But in fact, demolition had already begun. Another meeting ensued on January 3, 1988, at which the Minskoff engineers agreed to follow even more specific procedures, although there was still no written agreement.

Another week passed when Joe, visiting the museum on Sunday, was alarmed by what he saw when inspecting the site. The portion of the garage building that abutted the Skidmore House had been demolished, exposing the wall of the landmarked building. What Joe noticed was that the wall of the Skidmore House had already been backfilled without having been treated for waterproofing. Afraid the Old Merchants House might be subjected to the same destructive procedure in spite of Minskoff's assurances, Joe wrote Gene Norman, chairman of the Landmarks Preservation Commission, asking for immediate supervision to protect the Old Merchants House wall.

To clarify what would be necessary regarding the wall below ground level, Minskoff directed their consulting engineer to inspect the Old Merchants House party wall below grade. According to his report, the underground retaining wall Joe had wanted them to construct was not necessary, and waterproofing alone would provide further protection against water seepage. Minskoff informed Joe on January 27 that they would therefore proceed to backfill the wall after waterproofing measures were undertaken. No below grade retaining wall would be built.

Joe was furious, and on January 30, when he found that workers had carelessly removed façade bricks with a hammer rather than carefully separating them with a masonry saw as they had promised, he sent a desperate telegram to the engineer in charge and a copy to the Department of Buildings conveying the urgency of the work being satisfactorily performed under the supervision of the Department of Buildings and the Landmarks Commission.

Joe met on-site with the head of demolition and Minskoff's consulting engineer on February 3, 1988. They carefully inspected the building, and Joe insisted on seven specific actions necessary to protect the party wall and to make it structurally sound. After further consultations and meetings with the Landmarks Commission over the next three weeks, the Minskoff representatives eventually agreed to comply with Joe's demands.

On February 22, Joe met with the Minskoff attorney and engineer, the owners of the Skidmore House, and representatives from the Landmarks Preservation Commission. The next day he wrote Spencer Davidson:

> All is proceeding favorably for us at the Old Merchants House in the demolition of the former adjoining Hittner property. As a result we will have in writing and in drawing the construction details which are to reinforce our party wall structurally and make it as weatherproof as possible.

Joe was certain now that in spite of the contentious and prolonged negotiations with Minskoff, they had finally reached an understanding, and the forthcoming written details would be satisfactory. On March 11, a copy of the specifications for the work to be undertaken arrived in the mail for his approval.

Three days later in the middle of the night of March 14, 1988, Joe Roberto died in his sleep from a massive heart attack. The specifications were still laid out on his desk.

21

1988
A Legacy Imperiled

47. Joseph Roberto, 1908-88
The New York University Archives

April 20, 1988, would have been Joe Roberto's eightieth birthday. On that day, friends and colleagues gathered in the parlors of the Old Merchants House to participate in an affirmation of his life. There were flowers and sushi and wine and music, and many rose to speak of Joe's manifold accomplishments.

He had worked diligently to the very end at a profession he loved. And because he was so energetic, so involved with the Old Merchants House on a daily basis, his death came as a profound shock.

Now it was up to others to follow through on the complicated situation next door, solve the recurring problems, and somehow secure the future of the museum.

Spencer Davidson and Peter Krulewitch arranged for a meeting with executives and attorneys representing the Minskoff organization. In that meeting, which took place on April 28, the Minskoff representatives assured Davidson and Krulevitch that they would honor the promise they had made to carefully carry out the remaining demolition work.

Aware of the difficulties Joe had in getting Minskoff to agree to his demands, Krulevitch insisted that Minskoff put a declaration of their responsibility in writing. On May 2, 1988, a hand-delivered letter from Robert Kandel, the Minskoff attorney, to Leonard Carter, the attorney representing the Old Merchants House, spelled out their intention:

> We agree to continue to do the remaining demolition work and the subsequent demolition clearing and paving work with care and in a workmanlike manner so as to avoid any damage to the Old Merchant's House.
>
> If any of this work, however, results in any damage whatsoever to the Old Merchant's House, we also agree to assume at our sole cost and expense, full responsibility for any and all repair work required to restore the house to its present condition.

Less than two weeks later, in violation of all agreements, the Minskoff organization broke their promise in a brutal and shocking manner. Those who were at the museum on the afternoon of May 10, 1988, remember being startled by a sudden shattering explosion that caused the house to tremble. Certain an earthquake was in progress, they rushed to the door to discover a bulldozer was in the process of pushing the entire first story of the adjoining building, as well as the construction shed, into the street.

22

1988–97
The Realization of a Dream

On May 11, 1988, immediately after the outrageous demolition of the Hittner garage building, Max Bier and Liam O'Hanlon, the architect and engineer representing the Old Merchants House, met with Robert Redlien, consulting engineer for the Minskoff organization and Paul Pucci, who was in charge of the construction. They demanded that steps be taken to repair the damage and to protect the east wall of the museum now that it was exposed to water infiltration and the full force of the wind. Also present at this emergency meeting were Laura Alaimo from the Landmarks Preservation Commission, a representative from Community Board #2, Carol Roberto, and Frederick Brokaw, a museum trustee. At the end of the meeting, the Minskoff representatives had agreed to take the steps necessary to make the building sound and to immediately deliver in writing and drawings the specifications for the procedures they intended to follow.

But the specifications that arrived in engineer O'Hanlon's office the following week were not at all what he expected. He forwarded them to the Old Merchants House architect, Max Bier, pronouncing them "inconsistent, incomplete, inadequate, and irrelevant!" marginally noting the many inadequacies throughout.

This unsatisfactory response from the Minskoff organization was an indication the Old Merchants House was in for a protracted negotiation. Minskoff repeatedly balked at the requirements to make the necessary repairs as outlined in several engineering reports commissioned by the Old Merchants House, which specified the work necessary to repair the damage caused by the demolition and to protect the building from future damage. It would be almost a year and a half before the matter was resolved.

Finally, in October 1989, the Old Merchants House and the Minskoff organization reached an out-of-court settlement of $275,000. By this time, the damage resulting from the demolition was extensive. The house was in grave structural jeopardy. Lacking the support of the adjacent building for 17 months, the house had shifted. It stood like a house of cards, with the east and west sidewalls falling away from each other. If the walls were to separate from the floor joists altogether, a complete collapse would be inevitable. The front and rear façades were also affected. A crack resulting from the demolition had appeared at the southwest corner of the building causing portions of the woodwork of the interior window shutters to split open on two floors. Moreover, the interior walls and ornamental plaster had suffered extensive water damage.

Once again, to avoid collapse, the house faced a major intervention.

48. East Wall of the Old Merchant's House After Demolition
Once the adjacent building was demolished, the Old Merchants House was subject to destructive water damage and structural instability.

With Joe Roberto gone, executive direction was essential to guide the museum through this critical period. Aware of what could be lost forever, preservationists rallied around to help save the treasured historic site. Grants from private foundations and individuals provided funding for the first paid professional administration of the museum. In the fall of 1988, Janine Veto, a fundraising and arts management consultant, was hired as a part-time interim director, and in February 1990, Margaret Halsey Gardiner, who had been a board member since 1988 and board president since July 1989, was appointed to the permanent position of executive director. Her work was cut out for her. The house was perilously close to the point of no return, and the cost of restoring it would far exceed the financial resources available.

49. Upper Stories of East Wall After Demolition

But what could have been the end of the dream of preserving the Old Merchants House for future generations was actually the beginning of a new period of revitalization that continues to the present day.

Before the demolition of the adjacent building took place in May 1988, the restored museum had been open to the public for eight years. It had garnered the attention of preservationists throughout the country. A New York City landmark inside and out, a National Historic landmark, and the object of a nine-year restoration unparalleled for authenticity, the Old Merchants House was widely regarded as an invaluable historic document.

In what would be the first of many grants and offers of support over the next several years, a large grant of $173,000 was forthcoming in 1989 from the State of New York, made possible by the Environmental Quality Bond Act (EQBA) of 1986, which allotted some funds for the protection of historic properties. That grant plus the $275,000 Minskoff settlement provided the seed money to get the restoration started. The firm of Jan Hird Pokorny Associates was hired to undertake the necessary work.

In July 1990, two years after the demolition of the adjacent building at 31 East Fourth Street, the museum was closed until October for emergency repairs to the east and west walls. The crack at the southwest corner, which threatened the structural stability of the building, was repaired, and after excavation along the east wall, a below-grade drainage system was installed. Because water was leaking through the brick joints and open joist pockets, the east wall underwent the installation of a waterproofing membrane system.

By the spring of 1991, an architectural committee set up by the board had developed a five-phase plan to stabilize the building envelope and repair interior damage. The Pokorny firm would prepare a historic structures report, a requirement of the EQBA grant. The report when completed would examine the history of the house, the original construction and alterations to the building over the years, and the current condition of the building. It also would set forth a master plan for addressing the long-term restoration needs.

On March 4, 1991, the house was closed again, and the entire collection of furniture, textiles, and decorative arts was packed and stored as protection from the dirt and dust about to be stirred up. Professional art handlers were called in to pack the furniture and decorative arts, and the museum staff worked with the Fashion Institute of Technology, which provided nine graduate students in the museum studies program who would catalogue, photograph, and pack the almost 1,300 textiles.

Exterior work included masonry repair of the entire east wall above the foundation, rebuilding the southeast corner of the east wall; repair of two chimneys and chimney caps; and west wall masonry work, including repair of a large crack at the rear façade. To keep the walls from falling away from the building, 43 east-west and north-south structural ties, over double the original estimate, were used to secure the party walls to the floor joists at each level. Other interior work included repairs to floor boards and window enframements and shutter pockets.

In December 1991, ten months later, work remained to be done on the bedroom floors, but the ground and parlor floors were finally reopened to the public in time for the annual members' holiday party. So many wanted to celebrate the restoration, the party was held on two consecutive evenings. The bedrooms on the second floor were reopened in March 1992.

However, just as the house reopened, inspection of the building revealed that water seepage from the roof parapets over many years had severely damaged the plaster finish of rooms on the third floor, used for offices, and the fourth floor servants hall, used for collection storage. The exterior coating of Thoroseal on the west wall, which Joe Roberto had applied in 1972, was actually preventing the moisture in the saturated bricks from migrating to the outside. Water was instead seeping into the building. And to add to the problem, the mortar bonding the inner and outer wythes of brick was found to be no longer sound. The seeping water had removed much of the lime component; the wall was held together by only sand and gravity.

Repairs were undertaken in several phases, as funds became available. During the summer of 1992, the northeast section of the west parapet wall was reconstructed and new protective metal flashing installed to prevent rainwater from entering through the mortar joints.

When workers began removing the Thoroseal application from the west wall, they discovered it was not possible without damaging the brick. The solution was to remove the bricks and turn them around so that the Thoroseal was now on the inner face of the brick. The bricks were laid in a new softer, more moisture-transmissive mortar, and a path was thus created for moisture to move from, not into, the building.

In addition, a lower portion of the west party wall was rebuilt and new stainless steel anchors installed to provide additional support between the west wall and the rear façade. Significant masonry repairs were implemented at the east wall in 1992 as well, including re-pointing and replacement of brick masonry, and excavation and installation of a new higher performance system of waterproofing below grade.

Throughout the struggle of the 1990s and the challenge of moving forward towards a new century, the museum turned time and time again to government agencies, private foundations, corporate donors and individuals who generously funded these repairs and general expenses for operation.

And in December 1997, the dream of the founder, George Chapman, came true when the Vincent Astor Foundation awarded the museum a $1 million grant, contingent upon the museum raising $500,000. A total of $650,000 was raised, thus establishing an endowment fund, securing the financial stability of the museum into the future. For close to ten years after the Merchant's House received the Astor grant, structural restoration work continued. Finally, the building envelope was stabilized, and today the architecture firm of Jan Hird Pokorny Associates, Inc. carefully monitors the condition of the building, undertaking restoration projects when necessary.

Taking inflation into consideration, the $1 million grant was very close to the $100,000 George Chapman first envisioned raising in 1936. It just took a little longer than he thought.

18394

THE VINCENT ASTOR FOUNDATION
405 PARK AVENUE, NEW YORK, NY 10022

THE BANK OF NEW YORK

PAY TO THE ORDER OF Old Merchant's House

DATE December 1, 1997
AMOUNT $1,000,000.00

THE VINCENT ASTOR FOUNDATION

018394

50. The One Million Dollar Check from the Vincent Astor Foundation

In 1996, the name of the Old Merchants House was changed to
The Merchant's House Museum.

Acknowledgments

Throughout its history, every time the museum has faced an existential crisis, just the right person has come forward to save it. I am deeply grateful to each of them, for without their initiative, today there would be no house to write about.

To Pi Gardiner, the current executive director of the museum, I owe special thanks. Not only did she lead the house through the crisis brought about by the irresponsible demolition of the building next door, she has provided me with wise personal guidance throughout the years I have spent writing this history of the house we both love.

To George Chapman, Cornelia Van Siclen, and Joseph Roberto, I am grateful for their taking care to preserve records of their stewardship: correspondence, bills, contracts, grant proposals, budgets, minutes, a work log, an occasional handwritten record of a telephone conversation, or a "memo to self."

I am indebted to the following persons who were eyewitnesses to some part of the history of the museum and who kindly shared their memories with me:

Charles Fraser and Emmanuela Bricetti, who were George Chapman's nurses, and James Woods, an early museum board member.

Janet Hutchinson, who with her friend, Emmeline Paige, bridged a crucial one-year gap in the management of the museum.

Thanks to those who worked alongside Joe Roberto and remember him well: John Sanguiliano, Anita Brandt, Anthony Bellov, David Flaharty, and Clarice Watson, who was housekeeper for Joseph Roberto, worked for him at the museum during the restoration then continued as the museum housekeeper until her recent retirement. Her leaving marks the end of an era; she will be sorely missed.

Michael Devonshire of Jan Hird Pokorny Associates, who led the restoration of the nineties, was helpful in clarifying some of the more technical aspects of that restoration. And Ann Haddad discovered difficult-to-find photographs that helped tell the story. Thanks to them both.

I am grateful to Victoria Hagan, president of The Decorators Club, who made the archives of the club available to me, and to Lynne Kerr of her office, who helped me access them. A special thanks to Emily Eerdmans, for introducing me to her fellow club member, Anita Welch, who was on hand for the restoration of the 70s and who has proved to be a good friend as well as a valuable resource.

Mary Knapp
New York City

Albion, Robert Greenhalgh. *The Rise of New York Port [1815-1860]*. 1939. Reprint. Boston: Northeastern University Press. 1984.

Burnham, Alan, ed. *New York Landmarks: A Study & Index of Architecturally Notable Structures in Greater New York*. Middleton, CT: Wesleyan University Press. 1963.

Dunlap, David. "Planned Demolition of 3 Buildings Threatens Old Merchant's House." *The New York Times,* December 25, 1987.

Erskine, Helen Worden. *Out of This World.* New York: Putnam, 1953.

Hamlin, Talbot. *Greek Revival Architecture in America.* London: Oxford University Press. 1947.

Hayes, Helen and Anita Loos. *Twice Over Lightly.* New York: Harcourt Brace Jovanovich, Inc. 1972.

Horlick, Alan Stanley. *Country Boys and Merchant Princes.* Louisburg, PA: Bucknell University Press, 1975.

Hosmer, Charles Jr., *Presence of the Past: A History of the Preservation Movement in the United States Before Williamsburg.* New York: G.P. Putnam's Sons, 1965. See Ch. X, "William Sumner Appleton and the Society for the Preservation of New England Antiquities" for a summary of Appleton's work at SPNEA.

Huxtable, Ada Louise. "1832 Village Landmark Faces Demolition." *The New York Times,* February 18, 1965.

______. "A Funny Roll of the Dice." *The New York Times,* December 17, 1970.

______. "The Old Lady of 29 East Fourth St." *The New York Times,* June 18, 1972.

______. "A Landmark House Survives the Odds." *The New York Times,* February 28, 1980.

Jan Hird Pokorny, Architects & Planners. *Old Merchant's House: Historic Structure Report.* June, 1993.

Jones, L. Irwin. "The Old Merchant's House," 1936. U.S. Works Progress Administration, Architectural Section, New York City Unit. Index of American Design.

Lockwood, Charles. *Bricks and Brownstone.* New York: McGraw-Hill Book Company, 1972.

______. *Manhattan Moves Uptown: An Illustrated History.* Boston: Houghton Mifflin, 1976. A detailed description of the Bond Street neighborhood during the 1820s and 40s will be found in Ch. 4.

Schuyler, Montgomery. "The Small City House in New York." *The Architectural Record* 8, no. 4 (April-June, 1899).

Sharp, Lewis Inman. "The Old Merchant's House: An 1831/32 New York Row House." M.A. thesis, University of Delaware, 1968.

Williams, Roger M. "Landmark on East Fourth Street." *Americana* 9 (September/October 1981): 66–72.

Wood, Anthony C. "Pioneers of Preservation: An Interview with the Late Geoffrey Platt, the First Chairman of the Landmarks Commission." *Village Views* 4, no. 1 (winter 1987): 3–37.

______. "Pioneers of Preservation: An Interview with Harmon Goldstone, the Second Chairman of the Landmarks Preservation Commission." *Village Views* 4, no. 3 (summer 1988): 9–42.

Zucker, Howard. "Howard Zucker at the Old Merchant's House." *Victorian Homes* (winter 1986): 66–7.

Index

Made in the USA
Middletown, DE
25 March 2025